T0255669

Deep Belief Nets in C++ and CUDA C: Volume 2

Autoencoding in the Complex Domain

Timothy Masters

Apress®

Deep Belief Nets in C++ and CUDA C: Volume 2: Autoencoding in the Complex Domain

Timothy Masters
Ithaca, New York, USA

ISBN-13 (pbk): 978-1-4842-3645-1 ISBN-13 (electronic): 978-1-4842-3646-8
https://doi.org/10.1007/978-1-4842-3646-8

Library of Congress Control Number: 2018940161

Managing Director, Apress Media LLC: Welmoed Spahr
Acquisitions Editor: Steve Anglin
Development Editor: Matthew Moodie
Coordinating Editor: Mark Powers

Cover designed by eStudioCalamar

Cover image designed by Freepik (www.freepik.com)

Distributed to the book trade worldwide by Springer Science+Business Media New York, 233 Spring Street, 6th Floor, New York, NY 10013. Phone 1-800-SPRINGER, fax (201) 348-4505, e-mail orders-ny@springer-sbm.com, or visit www.springeronline.com. Apress Media, LLC is a California LLC and the sole member (owner) is Springer Science + Business Media Finance Inc (SSBM Finance Inc). SSBM Finance Inc is a **Delaware** corporation.

For information on translations, please e-mail editorial@apress.com; for reprint, paperback, or audio rights, please email bookpermissions@springernature.com.

Apress titles may be purchased in bulk for academic, corporate, or promotional use. eBook versions and licenses are also available for most titles. For more information, reference our Print and eBook Bulk Sales web page at www.apress.com/bulk-sales.

Any source code or other supplementary material referenced by the author in this book is available to readers on GitHub via the book's product page, located at www.apress.com/9781484236451. For more detailed information, please visit www.apress.com/source-code.

Printed on acid-free paper

Table of Contents

About the Author

Timothy Masters earned a PhD in mathematical statistics with a specialization in numerical computing in 1981. Since then he has continuously worked as an independent consultant for government and industry. His early research involved automated feature detection in high-altitude photographs while he developed applications for flood and drought prediction, detection of hidden missile silos, and identification of threatening military vehicles. Later he worked with medical researchers in the development of computer algorithms for distinguishing between benign and malignant cells in needle biopsies. For the last 20 years he has focused primarily on methods for evaluating automated financial market trading systems. He has authored the following books on practical applications of predictive modeling: *Deep Belief Nets in C++ and CUDA C: Volume 1* (Apress, 2018); *Assessing and Improving Prediction and Classification* (Apress, 2018); *Data Mining Algorithms in C++* (Apress, 2018); *Neural, Novel, and Hybrid Algorithms for Time Series Prediction* (Wiley, 1995); *Advanced Algorithms for Neural Networks* (Wiley, 1995); *Signal and Image Processing with Neural Networks* (Wiley, 1994); and *Practical Neural Network Recipes in C++* (Academic Press, 1993).

About the Technical Reviewer

Chinmaya Patnayak is an embedded software developer at NVIDIA and is skilled in C++, CUDA, deep learning, Linux, and file systems. He has been a speaker and instructor for deep learning at various major technology events across India. Chinmaya earned a master's degree in physics and a bachelor's degree in electrical and electronics engineering from BITS Pilani. He previously worked with the Defense Research and Development Organization (DRDO) on encryption algorithms for video streams. His current interest lies in neural networks for image segmentation and applications in biomedical research and self-driving cars. Find more about him at `chinmayapatnayak.github.io`.

Introduction

This book is a continuation of Volume I of this series. Extensive references are made to material in that volume. For this reason, it is strongly suggested that you be at least somewhat familiar with the material in Volume I.

All techniques presented in this book are given modest mathematical justification, including the equations relevant to algorithms. However, it is not necessary for you to understand the mathematics behind these algorithms. Therefore, no mathematical background beyond basic algebra is necessary to understand the material in this book.

However, the two main purposes of this book are to present important deep learning and data preprocessing algorithms in thorough detail and to guide programmers in the correct and efficient programming of these net algorithms. For implementations that do not use CUDA processing, the language used here is what is sometimes called *enhanced C*, which is basically C with some of the most useful aspects of C++ without getting into the full C++ paradigm. Strict C (except for CUDA extensions) is used for the CUDA algorithms. Thus, you should ideally be familiar with C and C++, although my hope is that the algorithms are sufficiently clear that they can be easily implemented in any language.

This book is roughly divided into four sections. Chapter 1 presents a technique for embedding class labels into a feature set in such a way that generative exemplars of the classes can be found. Chapters 2 and 3 present signal and image preprocessing techniques that provide effective inputs for deep belief nets. Special attention is given to preprocessing that produces complex-domain features. Chapter 4 discusses basic autoencoders, with emphasis on autoencoding entirely in the complex domain. This is particularly useful in many fields of signal and image processing. Chapter 5 is a reference for the DEEP program, available as a free download from my web site.

CHAPTER 1

Embedded Class Labels

A picture is worth a thousand words. Sometimes a lot more. In many applications, the ability to see what a classification model is seeing is invaluable. This is especially true when the model is processing signals or images, which by nature have a visual representation. If the developer can study examples of the features that the model is associating with each class, this lucky developer may be clued in to strengths and weaknesses of the model. In this chapter, we will see how this can be done.

The classification paradigm presented in Volume I of this deep belief net series involved greedily training one or more (usually more) RBM layers and then following these layers with one or more (usually one) feedforward layers trained with supervision, ultimately performing SoftMax classification.

A problem with this approach is that generative sampling produces random examples from the distribution of the entire dataset. This, of course, is useful. However, it would be even better if we could sample from the class distributions one at a time. This way we can see the features that the model associates with each individual class and thereby gain understanding of the model's internal operation.

This is easy to do. Simply stack two or more layers, greedily training each from the bottom up, as usual. But when one gets to the top RBM, append binary class labels to its input layer, which of course is the second layer from the top. This is illustrated in Figure 1-1 for a tiny network containing three layers and learning two classes.

T. Masters, *Deep Belief Nets in C++ and CUDA C: Volume 2*, https://doi.org/10.1007/978-1-4842-3646-8_1

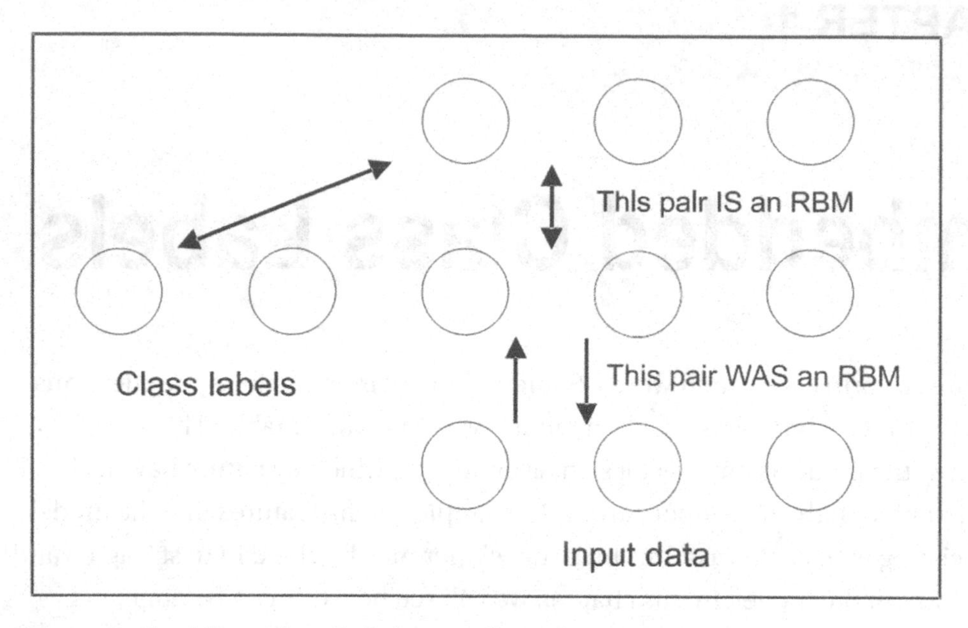

***Figure 1-1.** Embedding class labels in a three-layer network*

The bottom layer is the input data (just the predictors, ignoring class labels). Following the method of greedy training (Volume I, Chapter 4), we use an unsupervised algorithm to train the bottom two layers as an RBM. This RBM encapsulates the lowest-level features of the input space. If there are more than three layers, we continue the greedy training upward. But in this example there are just three layers.

Once the greedy training reaches the top-level RBM (the top two layers), we append to its input layer (the output of the prior layer) a neuron for each class. As each case is presented to the RBM during training, we set to 1-0 the neuron corresponding to the case's class, and we set the other neurons in the appendage to 0.0.

These two top layers are trained as an RBM in almost the usual way. The only catch is that the class-label appendage must remain a class indicator throughout the Gibbs sampling for contrastive divergence, with exactly one neuron at 1-0 and the others at 0.0. According to *A Fast Learning Algorithm for Deep Belief Nets* by Hinton et al. (2006), this can be accomplished by using Equation 1-1 to define the probability P_k of randomly selecting class k as the class to set to 1.0. In this equation, x_k is the net input of neuron k as given later in the chapter by Equation 1-5.

$$P_k = \frac{e^{x_k}}{\sum_i e^{x_i}} \tag{1-1}$$

To briefly review the basics of RBMs, let $\boldsymbol{v}$ be a vector of visible-layer activations. In the current context, this would be the activations of the prior (adjacent lower) layer, with the class labels appended. Then Equation 1-2 gives the activation probability vector of the corresponding hidden layer, the network's top layer in this context. Conversely, Equation 1-3 defines the visible activation probabilities corresponding to a given set of hidden neuron activations. In both of these equations, $f(.)$ is the logistic function given by Equation 1-4, applied element-wise. If this is not clear, see Chapter 3 of Volume I, which describes these concepts in great detail.

$$\boldsymbol{h} = f(\boldsymbol{c} + \boldsymbol{W}\boldsymbol{v}) \tag{1-2}$$

$$\boldsymbol{v} = f(\boldsymbol{b} + \boldsymbol{W}'\boldsymbol{h}) \tag{1-3}$$

$$f(t) = \frac{1}{1 + e^{-t}} \tag{1-4}$$

But for Equation 1-1 we need to break out the net inputs to the activation function. Given a set of hidden neuron activations, Equation 1-5 defines the values needed for that equation.

$$\boldsymbol{x} = \boldsymbol{b} + \boldsymbol{W}'\boldsymbol{h} \tag{1-5}$$

Recall that contrastive divergence learning performs one or more steps of Gibbs sampling by alternating Equations 1-2 and 1-3, always using random sampling to set the hidden layer activations and optionally (per the *mean field* option) using random sampling to set the visible layer activations. But even if mean field approximations are used (no sampling for the visible layer), we still randomly sample the neurons in the class-label appendage. In particular, exponentiate the visible layer net inputs (Equation 1-5) corresponding to the class labels and partition the probability according to Equation 1-1. The sum in the denominator of that equation is over the class labels only.

Code for Learning Embedded Labels

We are now ready to examine code for contrastive divergence learning, modified to handle embedded labels. It is strongly recommended that you review the corresponding section in Chapter 3 of Volume I, in which the nonembedded algorithm is discussed in detail. To avoid redundancy, the discussion here will speed over material already seen and focus on aspects of the code related specifically to handling embedded class labels. The calling parameter list for the threaded workhorse routine is as follows. It is almost identical to that shown in Volume I.

```
static void rbm2_threaded (
  int istart ,               // First case in this batch
  int istop ,                // One past last case
  int ncols ,                // Number of columns in data
  int n_inputs ,             // Number of inputs
  int embedded ,             // How many of the inputs are embedded class labels?
  double *data ,             // 'Training cases' rows by ncols columns of input data; 0-1
  int nhid ,                 // Number of hidden neurons
  int n_chain ,              // Length of Markov chain
  int mean_field ,           // Use mean field instead of random sampling?
  int greedy_mean_field ,    // Use mean field for greedy training?
  double *w ,                // Weight matrix, nhid sets of n_inputs weights
  double *in_bias ,          // Input bias vector
  double *hid_bias ,         // Hidden bias vector
  int *shuffle_index ,       // For addressing shuffled data
  double *visible1 ,         // Work vector n_inputs long
  double *visible2 ,         // Work vector n_inputs long
  double *hidden1 ,          // Work vector nhid long
  double *hidden2 ,          // Work vector nhid long
  double *hidden_act ,       // Work vector nhid long
  double *in_bias_grad ,     // Cumulate gradient here
  double *hid_bias_grad ,    // Cumulate gradient here
  double *w_grad ,           // Cumulate gradient here
  double *hid_on_frac ,      // Cumulate fraction of time each hidden neuron is on
  double *error              // Cumulates reconstruction criterion
  )
```

I'll now present the code in short segments, alternating with brief explanations. The first step is to initialize the random number generator seed (for sampling neuron activations) to a value that will change for each batch and each epoch. Then we zero the areas that will cumulate information across each batch.

```
{
  int k, randnum, icase, ivis, ihid, ichain ;
  double sum, *wptr, *dptr, P, Q, frand ;

  randnum = (istop + shuffle_index[0]) % IM ;
  if (randnum == 0) // Illegal value for this generator
    randnum = 1 ;

/*
  Zero the arrays that will cumulate gradient and error for this batch
*/

  for (ihid=0 ; ihid<nhid ; ihid++) {
    hid_bias_grad[ihid] = 0.0 ;
    hid_on_frac[ihid] = 0.0 ;
    for (ivis=0 ; ivis<n_inputs ; ivis++)
      w_grad[ihid*n_inputs+ivis] = 0.0 ;
    }

  for (ivis=0 ; ivis<n_inputs ; ivis++)
    in_bias_grad[ivis] = 0.0 ;

  *error = 0.0 ;
```

We loop over every case in this batch, with cases being randomly selected according to the shuffle index vector for the current epoch.

One small difference in this code relative to that in Volume I is accommodation for greedy training. Recall from the discussion in the first section of Chapter 4 that we may optionally sample the input of the layer being greedily trained. This was done in the greedy training code as of Volume I, but it has since been moved to the RBM routine, as shown in the following code. No big deal.

```
  for (icase=istart ; icase<istop ; icase++) {
    dptr = data + shuffle_index[icase] * ncols ;      // Point to this case in the data
    for (ivis=0 ; ivis<n_inputs ; ivis++)             // Get this case
      visible1[ivis] = dptr[ivis] ;
```

```
   if (! greedy_mean_field) {         // Optionally sample during greedy training
     for (ivis=0 ; ivis<n_inputs ; ivis++) {
       k = randnum / IQ ;
       randnum = IA * (randnum - k * IQ) - IR * k ;
       if (randnum < 0)
         randnum += IM ;
       frand = AM * randnum ;
       visible1[ivis] = (frand < visible1[ivis])  ? 1.0 : 0.0 ;
       }
     }
```

Compute the activation of each hidden neuron, and compute the reconstruction error if the compiler flag is set to do so in the slow but accurate way. In practice, this should usually be avoided because accuracy in the reconstruction error is not crucial and the extra work involved doing it here can be quite time-consuming.

```
   for (ihid=0 ; ihid<nhid ; ihid++) {
     wptr = w + ihid * n_inputs ;       // Weight vector for this neuron
     sum = hid_bias[ihid] ;
     for (ivis=0 ; ivis<n_inputs ; ivis++)
       sum += wptr[ivis] * visible1[ivis] ;
     Q = 1.0 / (1.0 + exp(-sum)) ;       // Probability
     hidden1[ihid] = hidden2[ihid] = Q ; // We'll need hidden2 for CD-k loop below
     hid_on_frac[ihid] += Q ;           // Need this for sparsity penalty
     }
#if RECON_ERR_DIRECT
   // Compute the reconstruction error the deterministic but expensive way
   for (ivis=0 ; ivis<n_inputs ; ivis++) {
     sum = in_bias[ivis] ;
     for (ihid=0 ; ihid<nhid ; ihid++)
       sum += w[ihid*n_inputs+ivis] * hidden1[ihid] ;
     P = 1.0 / (1.0 + exp(-sum)) ;
#if RECON_ERR_XENT
     *error -= visible1[ivis] * log(P+1.e-10) + (1.0 - visible1[ivis]) * log(1.0-P+1.e-10) ;
```

```
#else
      double diff = visible1[ivis] - P ;
      *error += diff * diff ;
#endif
      }
#endif
```

We now embark on the iterative Gibbs sampling from the Markov chain. The first step in each pass is to perform the mandatory random sampling of the hidden neuron activations.

```
   for (ichain=0 ; ichain<n_chain ; ichain++) {

      // Sample Q[h|x] to get next (binary) hidden layer.

      for (ihid=0 ; ihid<nhid ; ihid++) {
         k = randnum / IQ ;
         randnum = IA * (randnum - k * IQ) - IR * k ;
         if (randnum < 0)
            randnum += IM ;
         frand = AM * randnum ;
         hidden_act[ihid] = (frand < hidden2[ihid]) ? 1.0 : 0.0 ;
         }
```

Up to this point, except for the optional sampling for greedy training, this algorithm has been identical to the basic version shown in Volume I. But this is about to change because now we must accommodate the embedded class labels when we compute the visible layer.

In the code shown on the next page, we begin by using Equation 1-5 to compute the net input to a visible-layer neuron, exactly as in the "usual" case. But then, one of two things is true. Perhaps we are in the first part of the input vector, that encompassing the activations of the prior layer. In this case we proceed in the "usual" way. Or we are toward the end of the input vector, in the realm of the embedded class labels. In this case, we must prepare to use Equation 1-1. We compute reconstruction error here, making the arbitrary choice to not include class labels.

```
    for (ivis=0 ; ivis<n_inputs ; ivis++) {
      sum = in_bias[ivis] ;
      for (ihid=0 ; ihid<nhid ; ihid++)                    // Equation 1-5
        sum += w[ihid*n_inputs+ivis] * hidden_act[ihid] ;

      if (ivis >= n_inputs-embedded) { // If doing class labels, must set exactly one on
        if (sum > 300.0)               // Rare, but play it safe
          visible2[ivis] = exp ( 300.0 ) ;
        else
          visible2[ivis] = exp ( sum ) ;  // Prepare for Equation 1-1
        continue ;                     // Nothing more to do for the moment
        }                              // In embedded class labels section of input vector

      P = 1.0 / (1.0 + exp(-sum)) ;    // 'Usual' case; this is the probability

#if ! RECON_ERR_DIRECT
        // Compute the reconstruction error the stochastic but fast way
        // NOTE: If embedded labels, they are not included here!
        // This may or may not be a good thing.

        if (ichain == 0) {
#if RECON_ERR_XENT
          *error -= visible1[ivis] * log(P+1.e-10) + (1.0-visible1[ivis]) * log(1.0-P+1.e-10) ;
#else
          double diff = visible1[ivis] - P ;
          *error += diff * diff ;
#endif
          }
#endif
         if (mean_field)
           visible2[ivis] = P ;
         else {
           k = randnum / IQ ;
           randnum = IA * (randnum - k * IQ) - IR * k ;
           if (randnum < 0)
             randnum += IM ;
```

```
      frand = AM * randnum ;
      visible2[ivis] = (frand < P) ? 1.0 : 0.0 ; // Sample the activation
      }
   } // For each visible neuron, compute its probability and sample if not mean_field
```

At this point we have computed values for every neuron in the input layer. For those neurons early in the input vector, the outputs of the prior layer, these values are either the probabilities (if mean field approximation is used) or binary 0/1 samples taken according to the probabilities (if no mean field approximation). For the end of the input vector, those elements corresponding to the embedded class labels, the values are *e* raised to the power of the net produced by the hidden layer, as shown in Equation 1-5.

Now we have to use Equation 1-1 to define the probability P_k of randomly selecting class *k* as the class to set to 1.0, for each class *k*. This is done in the code shown here. An explanation follows.

```
if (embedded) {
  sum = 1.e-60 ; // Next two loops convert visible2 to cumulative probability
  for (ivis=n_inputs-embedded ; ivis<n_inputs ; ivis++)
    sum += visible2[ivis] ;        // Sum denominator of Equation 1-1

  for (ivis=n_inputs-embedded ; ivis<n_inputs ; ivis++) {
    visible2[ivis] /= sum ;                    // Convert to a probability
    if (ivis > n_inputs-embedded)              // After the first class, cumulate probability
      visible2[ivis] += visible2[ivis-1] ;
    }

  k = randnum / IQ ;
  randnum = IA * (randnum - k * IQ) - IR * k ;
  if (randnum < 0)
    randnum += IM ;
  frand = AM * randnum ;

  for (ivis=n_inputs-embedded ; ivis<n_inputs-1 ; ivis++) { // Find which to set to 1
    if (frand <= visible2[ivis])
      break ;
    }
```

```
    for (k=n_inputs-embedded ; k<n_inputs ; k++)
      visible2[k] = (k == ivis) ? 1.0 : 0.0 ;
    } // If embedded
```

The first step is to sum the denominator of Equation 1-1. Then we divide each term by their sum in order to get a probability. Sum these, element by element, to get a cumulative distribution function (*CDF*) and then generate a random number in the range 0-1. (This random generator is discussed in more detail in Volume I.)

The standard algorithm for selecting from among competing classes according to their probabilities is to pass through the cumulative distribution function and stop at the first bin for which the CDF equals or exceeds a uniform random number. This is implemented in the next loop. Finally, we set to 1.0 the visible neuron corresponding to the selected class, and we set all other neurons to 0.0.

Now we're back to the "usual" algorithm discussed in depth in Volume I. The Markov chain is completed by bouncing from the visible layer just computed back to the hidden layer.

```
for (ihid=0 ; ihid<nhid ; ihid++) {
  wptr = w + ihid * n_inputs ;      // Weight vector for this neuron
  sum = hid_bias[ihid] ;
  for (ivis=0 ; ivis<n_inputs ; ivis++)
    sum += wptr[ivis] * visible2[ivis] ;
  hidden2[ihid] = 1.0 / (1.0 + exp(-sum)) ;
  }
} // For Markov chain
```

This workhorse routine ends by computing the gradient for the weight matrix and bias vectors. We'll dispense with redundant explanations of this operation and just list the code for completeness.

```
for (ihid=0 ; ihid<nhid ; ihid++) {

  if (mean_field) {
    hid_bias_grad[ihid] += hidden1[ihid] - hidden2[ihid] ;
    for (ivis=0 ; ivis<n_inputs ; ivis++)
      w_grad[ihid*n_inputs+ivis] +=       hidden1[ihid] * visible1[ivis] -
                                          hidden2[ihid] * visible2[ivis] ;
    }
```

```
      else {
        k = randnum / IQ ;
        randnum = IA * (randnum - k * IQ) - IR * k ;
        if (randnum < 0)
          randnum += IM ;
        frand = AM * randnum ;
        hidden_act[ihid] = (frand < hidden1[ihid]) ? 1.0 : 0.0 ;
        hid_bias_grad[ihid] += hidden_act[ihid] - hidden2[ihid] ;
        for (ivis=0 ; ivis<n_inputs ; ivis++)
          w_grad[ihid*n_inputs+ivis] +=    hidden_act[ihid] * visible1[ivis] -
                                           hidden2[ihid] * visible2[ivis] ;
        }
      }

    for (ivis=0 ; ivis<n_inputs ; ivis++)
      in_bias_grad[ivis] += visible1[ivis] - visible2[ivis] ;

    } // For each case in this batch
}
```

Cross Entropy Reconstruction Error

The code just shown includes the compile-time option of computing reconstruction error using cross entropy, as defined in Equation 1-6, rather than the more common sum of squared errors.

$$\boldsymbol{ReconErr}_{XENT} = -\sum_{i}\left[x_i \log(v_i) + (1 - x_i)\log(1 - v_i)\right] \tag{1-6}$$

You may want to make the choice of computing the squared error versus cross entropy error a user choice, rather than hard-coding it into the program. Cross entropy is more meaningful when the inputs are strictly binary, or can be assumed to be probabilities, because cross entropy emphasizes extremely poor reconstruction (one term very near 1.0 while the other is very near 0.0). On the other hand, squared error has broader applicability and never is a poor choice, so I coded it this way, making the decision at compile time.

Fast vs. Slow Reconstruction Error Computation

In case the difference between the fast-and-accurate versus slow-and- approximate methods for computing reconstruction error is not clear, I ran a simple comparative test under a profiler. This test employed a simple stock price series (page 195) with a moving window of 252 days (one year of daily trades). An embedded model with 64 hidden neurons was trained for just one epoch. Figure 1-2 shows the time in milliseconds for each line using the fast method, while Figure 1-3 shows the times for the slow method. Notice how the inner-loop summation in the slow method is huge and in fact represents a sizable fraction of the total training time while serving no vital purpose.

```
        #if ! RECON_ERR_DIRECT
                    // Compute the reconstruction error the stochastic but fast way
                    // NOTE: If embedded labels, they are not included here!
25.94               if (ichain == 0) {
        #if RECON_ERR_XENT
                       *error -= visible1[ivis] * log(P+1.e-10) + (1.0-visible1[ivi:
        #else
26.20                  double diff = visible1[ivis] - P ;
26.12                  *error += diff * diff ;
        #endif
                       }
        #endif
```

Figure 1-2. *Reconstruction error computed with fast method*

```
             #if RECON_ERR_DIRECT
                    // Compute the reconstruction error the deterministic but expensive way
    0.10            for (ivis=0 ; ivis<n_inputs ; ivis++) {
   26.17               sum = in_bias[ivis] ;
   26.03               for (ihid=0 ; ihid<nhid ; ihid++)
1,690.40                  sum += w[ihid*n_inputs+ivis] * hidden1[ihid] ;
   27.74               P = 1.0 / (1.0 + exp(-sum)) ;
             #if RECON_ERR_XENT
                       *error -= visible1[ivis] * log(P+1.e-10) + (1.0 - visible1[ivis]) *
             #else
   26.19               double diff = visible1[ivis] - P ;
   26.28               *error += diff * diff ;
             #endif
   26.08               }
             #endif
```

Figure 1-3. *Reconstruction error computed with slow method*

Classifying Cases

When the model architecture has one or more unsupervised layers followed by one or more supervised layers, classification of a case is easy; simply propagate the case's predictor variables through the layers until the output layer is reached, and choose the class having highest output. But when the class labels are embedded along with input data, things become a lot more complicated.

The classic monograph *A Practical Guide to Training Restricted Boltzmann Machines* by Geoffrey Hinton (2010) outlines an efficient approach. The idea is that you present the case to the model once for each class. Each time, set a different class-label neuron on, leaving the others off. Whichever presentation has the lowest free energy is the chosen class.

There are numerous expressions for the free energy of a configuration, having various degrees of theoretical clarity and computational complexity. The most computationally efficient formula is given by Equation 1-8, with the x_j terms being the components of the vector defined in Equation 1-5. To ensure clarity, this term is also defined in Equation 1-7. In these equations, ***b*** is the hidden bias vector, and ***a*** is the visible bias vector.

$$\boldsymbol{x}_j = \boldsymbol{b}_j + \sum_i \boldsymbol{W}_{ji} \boldsymbol{v}_i \tag{1-7}$$

$$\boldsymbol{F}(\boldsymbol{v}) = -\sum_i \boldsymbol{v}_i \boldsymbol{a}_i - \sum_j \mathbf{log}\left(\mathbf{1} + \boldsymbol{e}^{\boldsymbol{x}_j}\right) \tag{1-8}$$

What makes this representation especially good for computation is the fact that much of the effort needs to be done only once, not repeated for each presentation. In nearly all practical applications, the number of predictors vastly exceeds the number of classes. Thus, the sum in Equation 1-7 can be done in advance across all predictors. Since for each trial class exactly one v_i is 1.0 and all others are 0.0, completing the sum involves just adding in a single element of the weight matrix.

Moreover, the first term of Equation 1-8 involves, for the most part, summing quantities (the predictor terms) that are the same for all trial classes. Thus, we can completely ignore that part of the computation. With all this in mind, here is a code fragment for performing the required calculations. An explanation follows.

```
n_layer_inputs = number of pre-class inputs to the last (top) layer
ntarg = number of classes
nhid = number of hidden neurons, indexed by j in Eq (1.8)
wtptr = weight matrix (nhid rows by n_layer_inputs+ntarg columns)
hbptr = hidden bias vector
vbptr = visible bias vector
```

```
for (icase=istart ; icase<istop ; icase++) {
  rptr = tmp_inputs + icase * max_neurons ; // Point to this case in local storage

  // Evaluate for each j the pre-class sum for xj of Equation 1-7

  for (j=0 ; j<nhid ; j++) {
    wptr = wtptr + j * (n_layer_inputs + ntarg) ; // Visible weights for this hidden neuron
    sum = hbptr[j] ;
    for (i=0 ; i<n_layer_inputs ; i++)
      sum += rptr[i] * wptr[i] ;
    rbm_work_vec1[j] = sum ;                    // Store Equation 1-7 partial sums here
    }

  // Evaluate F(v) (except for pre-class of first term) for each trial class

  for (itarg=0 ; itarg<ntarg ; itarg++) {
    crit = vbptr[n_layer_inputs+itarg] ;
    for (j=0 ; j<nhid ; j++) {
      wptr = wtptr + j * (n_layer_inputs + ntarg) ;
      crit += log ( 1.0 + exp ( rbm_work_vec1[j] + wptr[n_layer_inputs+itarg] ) ) ;
      }
    if (itarg == 0 || crit > largest) {
      ipred = itarg ;
      largest = crit ;
      }
    }
... Do something with chosen class (ipred) ...
  } // For all cases
```

As we begin to process a case, we point to its predictor vector with rptr. In the most general situation, this will be the activations of the layer prior to this top-layer RBM where the embedded class labels reside.

The first step then is to compute and save in rbm_work_vec1 the partial sums of Equation 1-7. These are the sums across all predictors, stopping where the embedded class labels begin. These sums, of course, will be identical for all subsequent trial classes and so are computed once and preserved for the class trials.

We then begin the trial class loop. Initialize crit to be the visible bias term for the neuron corresponding to the trial class. To understand this, look back at the first term of Equation 1-8. The components of this sum corresponding to the predictors are identical for all trial classes, so they are ignored. Also, every class except the trial class has its visible activation zero by definition. So, we are left with just a single component of this sum, the bias corresponding to the trial class.

The only expensive operation in this algorithm is summing over j (hidden neurons) the second term of Equation 1-8. In the prior step we precomputed and saved the sum of Equation 1-7 for the predictors. To complete the sum for a given hidden neuron and trial class, all we need to do is add the corresponding element of the weight matrix. (Recall that v_i is 1.0 for the trial class and 0.0 for all other classes.)

All that remains is to keep track of which trial class has the best criterion. Note that we minimize Equation 1-8, but in this computation we flipped the signs of both terms. Thus, we maximize the criterion computed in this code.

When the trial class loop is complete, ipred will identify the chosen class. Do whatever you want with this information.

Class-Conditional Generative Sampling

An important reason for performing classification by embedding class labels is to allow the developer to see examples of the features that the model has encapsulated for each individual class. We now explore how this is done.

The process of class-conditional generative sampling is extremely similar to that for unconditional sampling. That topic was presented in depth in Chapter 4 of Volume I, so to avoid redundancy we will completely skip much of the material that is in common. The difference is entirely in the Gibbs sampling from the Markov chain. The complete algorithm prior to this step includes first propagating upward through prior layers any user-supplied input before implementing the chain and then propagating back down the final result from the chain. These "pre" and "post" steps are omitted here for clarity. See Volume I for details on them. Also, complete source code for this algorithm is available for free download from my web site. We now focus exclusively on the Gibbs sampling, which is the core operation.

Recall that ordinary Gibbs sampling involves repeatedly bouncing back and forth between the layers. One computes hidden-layer probabilities from the visible layer and samples accordingly. Then one computes visible-layer probabilities from the hidden layer.

For generative sampling we avoid sampling in this direction, as it just introduces unnecessary randomness.

To modify this procedure for sampling from a fixed class, all we need to do is clamp the class-label neurons to their desired values, which would be 1.0 for the class being studied and 0.0 for all other classes. In other words, each time we move from the hidden layer to the visible layer, for the class-label neurons we ignore activations computed from the weight matrix and hidden activations (Equation 1-3. Instead, fix the visible activations for the class-label neurons to their binary values.

We now present a large code fragment for the Gibbs sampling, modified for class-conditional sampling. Explanations are interspersed.

The sampling chain employs several variables that are defined as follows:

> nin: *The number of inputs to this RBM layer, not counting class labels.*
>
> nhid: *The number of hidden neurons.*
>
> embedded: *The number of classes.*
>
> input_vis: *Did the user supply a sample starting input (versus random hidden)?*
>
> w: *Weight matrix for the RBM.*
>
> hbptr: *Hidden bias vector.*
>
> vbptr: *Visible bias vector.*
>
> vis_layer: *Visible layer activation vector.*
>
> hid_layer: *Hidden layer activation vector.*
>
> clamp: *Clamp to a single class (always true in this context)?*
>
> clamped_class: *Class index to clamp to.*

The first step in the chain is to compute hidden-layer activations from the visible layer.

```
for (ichain=0 ; ichain<nchain ; ichain++) {

  if (ichain || input_vis) {                    // Skip first vis-to-hid if random hidden
    for (ihid=0 ; ihid<nhid ; ihid++) {          // Visible to hidden, with sampling
      wptr = w + ihid * (nin + embedded) ;       // Weight vector for this neuron
```

```
sum = hbptr[ihid] ;                              // This hidden neuron's bias
for (ivis=0 ; ivis<nin ; ivis++)                  // Equation 1-2
   sum += wptr[ivis] * vis_layer[ivis] ;

if (embedded) {                                  // Always true in this context
   if (clamp)                                    // Usual situation
      sum += wptr[nin+clamped_class] ;
   else {                                        // Funny for user to embed but not clamp
      for (ivis=nin ; ivis<nin+embedded ; ivis++)
         sum += wptr[ivis] * vis_layer[ivis] ;
      }
   }

Q = 1.0 / (1.0 + exp(-sum)) ;
frand = Uniform random number in the range 0-1
hid_layer[ihid] = (frand < Q) ? 1.0 : 0.0 ;
}
}
```

The large block that makes up the start of the chain loop is executed only if we are either past the first iteration of the chain or the user has specified visible inputs to start the chain. Such an input would typically be a training case. In the more common case that the user specified random activations for the hidden layer, we obviously would not want to propagate undefined visible inputs the first time through the loop!

The weight matrix has nin+embedded columns, so we use this fact to point to the weight vector for the hidden neuron being computed.

The code shown handles both embeded and nonembedded architectures, so we need to have an if(embedded) check. In the current context this block will always be executed because we are talking about embedded architectures here, and embedded is the number of classes.

When the user is creating generative samples, said user will almost invariably decree that a specified class clamped_class is to be clamped on so that the samples are conditional on that class alone. Since the clamped class's neuron has an activation of 1.0 and all other label neurons have a value of 0.0, all we need to do to complete the sum for Equation 1-2 is add in the corresponding weight.

This routine does allow for the possibility that the model has an embedded architecture, but the user does not want to clamp a particular class. In this situation, generative sampling would be for the distribution of the entire training set, in other words, all classes, rather than being limited to a single class. When this is done and if the user is starting the chain with visible-layer inputs rather than random hidden neurons, then the caller of this routine must ensure that not only the predictor section of vis_layer is filled in but the class-label section as well. If the starting input is a training case, then it makes sense to initialize the label neurons for the class of that case. But actually, the initialization is not critical; even all zeroes would do. This is because as long as the chain uses enough iterations, Markov mixing will swamp out the initial values of the label neurons.

The final action in this step is to apply the logistic activation function and sample.

The next step in the Markov chain is to compute reconstructed visible activations from the hidden activations. We begin by computing the sum shown in Equation 1-3. As long as we are in the predictor section of the visible layer (the else clause next), we just apply the activation function to the sum, giving the activation probability. We do not sample, as this would introduce undesirable superfluous randomness.

But once we reach the class-label section of the visible layer (ivis >= nin), we need to pay attention. If the user is clamping, we are all done. Look back at the prior block of code, which does the visible-to-hidden computation, and note that if we are clamping, the label section of the visible vector is ignored. But if the user is not clamping so that all classes are represented in the generative sampling, then we need to prepare for the special case of random sampling in the label section. This will be handled in the next block of code.

```
for (ivis=0 ; ivis<nin+embedded ; ivis++) {          // Hidden to visible
  sum = vbptr[ivis] ;                                  // Visible bias
  for (ihid=0 ; ihid<nhid ; ihid++)                    // Equation 1-3
    sum += w[ihid*(nin+embedded)+ivis] * hid_layer[ihid] ;
  if (ivis >= nin) {        // If doing class labels, must set exactly one on
    if (clamp)              // If not clamping, labels set randomly per probability
      break ;               // But if clamping, nothing to do here
    if (sum > 300.0)        // Rare but be safe
      vis_layer[ivis] = exp ( 300.0 ) ;
```

```
      else
        vis_layer[ivis] = exp ( sum ) ;
      }
    else
      vis_layer[ivis] = 1.0 / (1.0 + exp(-sum)) ;
    }
```

We come now to the final block of code in the Markov chain loop for Gibbs sampling. For completeness it is shown on the next page. However, we will dispense with an explanation because this same algorithm was shown on page 9 and explained in detail on the following page. This code is executed only for embedded architectures with no clamping. In this situation, the chain iteration must reconstruct the label section of the visible layer using SoftMax sampling.

```
if (embedded && ! clamp) {      // No point doing it if we are just clamping
                                // Because claming ignores the label section
  sum = 1.e-60 ; // Next two loops convert vis_layer to cumulative probability
  for (ivis=nin ; ivis<nin+embedded ; ivis++)
    sum += vis_layer[ivis] ;

  for (ivis=nin ; ivis<nin+embedded ; ivis++) {
    vis_layer[ivis] /= sum ;
    if (ivis > nin) // After the first class, cumulate probability
      vis_layer[ivis] += vis_layer[ivis-1] ;
    }

  k = randnum / IQ ;
  randnum = IA * (randnum - k * IQ) - IR * k ;
  if (randnum < 0)
    randnum += IM ;
  frand = AM * randnum ;

  for (ivis=nin ; ivis<nin+embedded-1 ; ivis++) { // Find which one to set to 1
    if (frand <= vis_layer[ivis])
      break ;
    }
```

```
   for (k=nin ; k<nin+embedded ; k++)
      vis_layer[k] = (k == ivis) ? 1.0 : 0.0 ;
   } // If embedded and not clamped
```

This code handles sampled reconstruction of the label section of the visible layer. Please see page 9 and the subsequent explanation for a thorough discussion of this algorithm. Recall that there is no need to reconstruct the label section of the visible layer if we are clamping a class because in that situation the label section is ignored when computing the hidden layer.

CHAPTER 2

Signal Preprocessing

Untold volumes of material have been written on the topic of signal preprocessing for prediction and classification. The treatment here will not pretend to do more than lightly scratch the surface of this huge topic. In fact, we will not even present a survey of techniques. You are invited to peruse other references for specialized topics.

Rather, this chapter will present several specific methods for preprocessing time series in order to compute variables that make suitable inputs to deep belief nets (and, indeed, many other models). The methods shown here are simply those that I favor in my own work. The exclusion of commonly used techniques, or methods that are favorites of other developers, is not meant to imply that I find them inferior. In fact, my hope is to include additional algorithms in later volumes of the book and versions of the program.

The first method presented is the easiest: just use the raw values of the series, perhaps transformed with logs or differences or both, to predict the next value of the series. This doesn't sound very sophisticated, but in fact it can be enormously powerful, especially for deep belief nets that have exceptional ability to detect patterns.

The second method is slightly more complex. One defines a function of a time series. In the DEEP program. this function is the linear slope within a recent window over the series, although in general it can be anything. This function is then moved across the history. Sets of adjacent values of this function serve as inputs to the model. In this way, the model sees the "path" of this function across time. The rate at which the function changes its value can also be useful.

The third method involves computing the discrete Fourier transform of a moving window.

The fourth method computes Morlet wavelets in a moving window to identify the state and velocity of a quasiperiodic component that may appear and disappear in a time series.

The final method traces an XY path on a plane, computing a Fourier transform.

T. Masters, *Deep Belief Nets in C++ and CUDA C: Volume 2*, https://doi.org/10.1007/978-1-4842-3646-8_2

Simple, Minimal Transformation

The simplest approach to time-series processing with deep belief nets (or most other models) is to just use the raw or minimally transformed values of the series as they exist in a historical window.

Logs and Differences

Often, we will want to take a log of the series to handle the common situation of the standard deviation of the values being proportional to the values themselves. Or it may be that the values of the series have no predictive power, but the changes from sample to sample carry useful information. Finally, we may want to do both: take the log of each point in the series and then use the difference in these logs to serve as predictors as well as the target being predicted.

The classic example of taking logs and then differencing is when processing equity prices. Suppose that an equity in time period *A* has an average price that is ten times higher than that in time period *B*. Then, all else being equal, we would normally see the daily variation in time period *A* being ten times that in time period *B*. Thus, we should take logs of the prices in order to stabilize the daily variation across the entire historical extent of the series.

Moreover, the actual price of an equity says little about its recent behavior. Rather, it is the *change* in its price that conveys useful information. Thus, we need to compute predictors as the (log) price changes. This change is also a useful target; who doesn't want to know the expected price change from today to tomorrow?

This is a good time for a brief digression to discuss a common error made when computing changes in equity prices or any other series whose variation is proportional to its value. Developers of financial trading systems intuitively think of percent returns. If a price moves from 1000 to 1020, the percent return is 100 * (1020 - 1000) / 1000 = 2 percent. Hence, developers might be inclined to use this transformation instead of the difference of logs.

The problem with percent moves is that they are not symmetric. If we make 10 percent on a trade and then lose 10 percent on the next trade, we are not back where we started. If we score a move from 100 to 110 but then lose 10 percent of 110, we are at 99, not back at 100. This might not seem serious, but if we look at it from a different direction, we see why it can be a major problem. Suppose we have a long trade in which the market moves from 100 to 110, and our next trade moves back from 110 to 100. Our net equity change is zero. Yet we have recorded a gain of 10 percent, followed by a loss of 9.1 percent, for a net gain of almost 1 percent! If we are recording a string of trade returns for statistical analysis, these errors will add up fast, with the result that a completely worthless trading system can show an impressive net gain! This will invalidate almost any performance test.

There is a simple solution that is used by professional developers: convert all prices to the log of the price, and compute price changes or trade returns (targets in predictive modeling) as the difference of these logs. This solves all of the problems. For example, a trade that captures a market move from 10 to 11 is 2.39789 - 2.30258 = 0.09531, and a trade that scores a move from 100 to 110 is 4.70048 - 4.60517 = 0.09531. If a trade moves us back from 110 to 100, we lose 0.09531 for a net gain of zero. Perfect.

A nice side benefit of this method is that smallish log price changes, times 100, are nearly equal to the percent change. For example, moving from 100 to 101, a 1 percent change, compares to 100 * (4.61512 - 4.605) = 0.995. Even the 10 percent move shown earlier maps to 9.531 percent. For this reason, we will treat returns computed from logs as approximate percent returns.

Windows and Shifting

The usual way of generating training and testing data from a time series is by employing a moving window over which predictors and targets are computed. The user specifies a lookback period that encompasses all data needed to compute the predictors and the target. For example, suppose we want to use the 10 most recent values of the series, perhaps transformed as described in the prior section, to predict the next (possibly similarly transformed) value of the series. If we are not differencing, we would need a window 11 cases long: the 10 predictors and the target. However, if we are working with differences, as would be the case if our data is equity prices, we would need a window containing 12 cases in order to have 11 differences.

The user must also specify the number of observations to shift forward to get the next training/testing case. Usually we would shift the window forward just one observation at a time because this provides the greatest number of cases, always an admirable goal. However, there are two reasons why we might need to shift forward more than one observation. It may (rarely!) be the situation that we simply have too much data, and the number of cases generated with single-observation shifts would be overwhelming, requiring too much computer time to process. I've never been so lucky, but you never know.

A more common and serious reason for multiple-observation shifts is if the source series has significant serial correlation. This is especially important if we are generating a test series. If the targets in a test situation have serial correlation, the effect on all conventional statistical tests is to decrease the degrees of freedom of the test, resulting in overly optimistic test results. This is discussed more fully in most Statistics 101-level textbooks, which emphasize the importance of having independent observations.

A more subtle but potentially deadly effect of serial correlation in generated test data is when the generated cases are split into a training set and a test set. Consider what happens at the boundary where the training data ends and the test data begins. There will be cases near the end of the training period whose predictors *and* targets are correlated with those of cases near the beginning of the test period. This is a form of *future leak*, a phenomenon in which information about the future test period leaks into the training set. The end result is that optimization of a predictive or classification model using this training data will inadvertently also tend to optimize performance in this section of the test data, again resulting in unduly optimistic results for a completely different but equally dangerous reason.

Thus, if the source series has serial correlation, ***the window shift must be great enough to bypass all serial correlation***. This is vital.

Pseudocode for Simple Series Processing

In this section we present a highly abbreviated algorithm in C-like pseudocode for implementing the ideas discussed in the prior two sections. You can find this "code" in the file SERIES.TXT. The following quantities are specified by the user:

length: Number of historical predictors desired

shift: Number of observations to shift window for each generated case

nature: One of the following four values:

- RAW: Use the actual observations
- RAWLOG: Use the log of the actual observations
- DIFF: Use the difference of consecutive observations
- DIFFLOG: Use the difference of the logs of consecutive observations

The required number of cases in the source series moving window is length, plus one extra for the target (which is the next observation after the predictors), plus one more if we are differencing. Compute the number of extra observations needed and then initialize for the processing loop.

```
if (nature == RAW || nature == RAWLOG) // No differencing?
  extra = 1 ;       // Just the target
else                // But if differencing
  extra = 2 ;       // We need to include one more

window = [Allocate memory for length+extra observations] ;

n_in_window = 0 ;          // Must fill window before we start generating database cases
shift_count = shift-1 ;    // Counter for window shifting
n_cases = 0 ;              // Will count generated cases

Initialize 'Most recent value of source series' to the first observation
```

The main processing loop is here. We just initialized the "most recent" value of the source series to be the first observation.

```
for all cases {

  x = [Most recent value of source series] ;
```

The following logic handles the initial filling of the window. We cannot begin computing output values until the window is full.

```
  if (n_in_window == length+extra) { // If window is full, shift down and put new one in
    for (i=1 ; i<n_in_window ; i++)
      window[i-1] = window[i] ;
    window[n_in_window-1] = x ;
    }

  else
    window[n_in_window++] = x ;    // Keep filling the window

  if (n_in_window < length+extra)    // Do nothing if it's not full yet
    continue ;
```

When we get here, the window is full, so we are ready to compute the variables for this window. But check how much the user wants us to shift the window between variables for the database.

```
if (++shift_count < shift)
  continue ;
shift_count = 0 ;
```

We can now compute the output record containing the predictors and the target. The target is the last item placed in the output record.

```
if (nature == RAW) {
  for (i=0 ; i<=length ; i++) // Includes target
    record[i] = window[i] ;
  }

else if (nature == RAWLOG) {
  for (i=0 ; i<=length ; i++) {
    if (window[i] > 0.0)
      record[i] = log ( window[i] ) ;
    else
      record[i] = -1.e60 ;    // Arbitrary choice for illegal situation
    }
  }

else if (nature == DIFF) {
  for (i=0 ; i<=length ; i++) // Includes target
    record[i] = window[i+1] - window[i] ;
  }

else if (nature == DIFFLOG) {
  for (i=0 ; i<=length ; i++) { // Includes target
    if (window[i+1] > 0.0)
      x = log ( window[i+1] ) ;
    else
      x = -1.e60 ;              // Arbitrary choice for illegal situation
```

```
      if (window[i] > 0.0)
        record[i] = x - log(window[i]) ;
      else
        record[i] = x + 1.e60 ;   // Arbitrary choice for illegal situation
      }
   }

Save this record to the database

Advance source series to next observation

++n_cases ;    // Count cases generated
} // For all cases
```

Tail Trimming

Most predictive and classification models, especially including deep belief nets, behave badly when presented with outliers, cases that have a predictor variable or target with a value that is far outside the normal range for the variable. The model will tend to assign undue importance to an outlier case, treating it as something very special, to the detriment of the other more "normal" cases. Thus, it is often in our best interest to pass through the database and trim the tails of every predictor as well as the target. There are some highly sophisticated algorithms for doing this, but the simple method shown here works well most of the time. The user specifics a small fraction trim, typically 0.01 to 0.1 or so, of each tail that is to be removed. The pseudocode for doing this is as follows:

```
for (ivar=0 ; ivar<=length ; ivar++) {          // For every predictor, as well as target
  for (i=0 ; i<n_cases ; i++)                   // Copy this data record to a work vector
    work[i] = database[i*n_vars+ivar] ;         // Database has n_vars columns
  qsortd ( 0 , n_cases-1 , work ) ;             // Sort work vector ascending
  k = (int) (trim * (n_cases+1)) - 1 ;          // Unbiased quantile index
  if (k < 0)                                    // Protect against user setting trim tiny
    k = 0 ;
  xmin = work[k] ;                              // Lower bound for trimmed data
  xmax = work[n_cases-1-k] ;                    // And upper bound
```

```
   for (i=0 ; i<n_cases ; i++) {                // Process all cases for this variable
      if (database[i*n_vars+ivar] < xmin)   // Trim left tail
         database[i*n_vars+ivar] = xmin ;
      if (database[i*n_vars+ivar] > xmax)  // trim right tail
         database[i*n_vars+ivar] = xmax ;
      }
   } // For all predictors
```

This code processes each variable individually. All cases of the variable are copied to a work vector and sorted ascending. We compute the unbiased index of the value corresponding to the user-specified trim fraction, which in turn determines the lower and upper thresholds for trimming outliers. Finally, we pass through the database, replacing values beyond these thresholds with trimmed values.

If you are interested in a far more sophisticated method for tail trimming, one that among other desirable properties retains ordinal relationships, you can find the algorithm and code in my book *Testing and Tuning Market Trading Systems.*

Example of Simple Series Creation

Here is an example of how we can create predictors and targets from a simple univariate series. The source series is prices of the S&P 100 Index OEX. Because this is an equity price series, our predictors and target are computed as the difference of log prices. The DEEP.LOG file will be discussed in greater detail in Chapter 6, so here we will focus exclusively on the content of the file as it relates to generating a simple series of predictor and target variables for model building. The log file records the user's choices.

Most of the user's choices shown in this Figure 2-1 log file excerpt are clear, as they have already been discussed. The only item of note is the statement "Above/below median class variables are generated and the model is a classifier." This is interesting in view of the fact that no model has been mentioned; we just read a market price series and generated a dataset from a moving window. This will become clearer in Chapter 6, but the short version of the story is that the user specified that the generated dataset may be used with a classification model, and the class of each case is determined by whether the target for that case is above versus at or below the median target across the entire dataset. The program thus computes the median and above/below flag variables for use later.

```
Reading series file D:\DEEP\TEST\OEX_TRAIN.TXT for 'simple' variables
   The log of the series is differenced to create the predictors and target
   Data is extracted from column 2
   Window length (number of predictors per case) is 20
   Shift (spacing of adjacent cases in series) is 1
   Tails are trimmed by 5.000 percent
   Targets are multiplied by 1.000
   Above/below median class variables are generated and model is classifier
   No header record is skipped

Simple series trimming is as follows...
       Variable      Old min        New min        New max        Old max

         Lag_19      -0.23540       -0.01747        0.01737        0.10151
         Lag_18      -0.23540       -0.01747        0.01737        0.10151
         Lag_17      -0.23540       -0.01747        0.01737        0.10151
          ...
          Lag_2      -0.23540       -0.01745        0.01734        0.10151
          Lag_1      -0.23540       -0.01745        0.01734        0.10151
          Lag_0      -0.23540       -0.01745        0.01734        0.10151
         Lead_1      -0.23540       -0.01745        0.01734        0.10151
```

Figure 2-1. *DEEP.LOG record of series generation, part 1*

Figure 2-1 shows the results after tail trimming. There are 20 predictors, labeled *Lag_0* (the current value) through *Lag_19*. The target is labeled *Lead_1* because it is the next value in the price series after the predictors. The columns in this table show the minimum and maximum of each variable, before and after trimming. Note that trimming just 5 percent of each tail produces a huge change in the min and max, an indication that the variables had very heavy tails.

Figure 2-2 continues the log file's report on the generation of this set of simple-series variables. Because the user requested that above/below median flags be generated, the median target is reported. Be aware that the median is preserved as part of the dataset. If a trained model exists at the time the series is read, the assumption is that we are now reading a test set, so the median is not recomputed. Rather, the existing median is used to compute the above/below flags, which prevents utilization of information that would not be known in real life.

The table in this figure shows basic statistics for the computed variables. Note the two additional variables at the end, *Lead_Pos* and *Lead_Neg*. We have two flag variables because most classification models, including those in this book, use a separate flag variable for each class. The flag for the correct class is set to 1.0, while that for all other classes is set to 0.0.

```
Simple series defined for training; median split = 0.00050

Basic statistics after any trimming...

        Variable        Mean          StdDev          Min          Max

          Lag_19      0.00036       0.00883     -0.01747      0.01737
          Lag_18      0.00036       0.00883     -0.01747      0.01737
          ...
           Lag_1      0.00036       0.00881     -0.01745      0.01734
           Lag_0      0.00036       0.00881     -0.01745      0.01734
          Lead_1      0.00036       0.00881     -0.01745      0.01734
        Lead_Pos      0.50000       0.50000      0.00000      1.00000
        Lead_Neg      0.50000       0.50000      0.00000      1.00000

Sucessfully read 7652 cases with 23 variables
```

Figure 2-2. *DEEP.LOG record of series generation, part 2*

Displaying Differenced Generative Samples

The concept of generative sampling was introduced in Volume I and extended to embedded models on page 15 of this book. It is easy to display generated representations of a time series when the predictors are values of the series itself. And dealing with logs in the display is not even needed in practice (at least in my experience) because any time the nature of the data is such that a log transform is needed to stabilize variation, a visual display of the series is more understandable in the log domain anyway. Direct display of generated changes is also straightforward. But when we have differenced the data, visual display of *pre-differenced* representations based on generated samples of *differences* is trickier.

Just to be clear, suppose we have defined the predictors (and target if we have a predictive model) to be the differences of a series (possibly after a log transform). It is usually most informative to display generated samples of the series itself rather than the differences. The human brain is poor at reconstructing a series by visually summing displayed differences. Samples generally make a lot more intuitive sense when they are in the pre-differenced domain. We will now present an effective way to do this.

Please keep in mind that in this discussion we will ignore log transforms; if the data was stabilized with a log transform, the visual display will remain in the log domain where, in most situations, it will be more interpretable.

The first issue to be dealt with is scaling. We want the displayed signals to correctly reflect the actual variation in the series; little movement in a series throughout its time window can be a legitimate characteristic. But if we were to scale all samples to occupy the full vertical range of the plot, we would lose this vital piece of information. For this reason we must pass through the entire training set and find the maximum range of the pre-differenced series. All displayed generative samples will be scaled relative to this range. The code fragment on the next page illustrates this simple procedure. In this code, inptr for each case points to the differences, the predictors for the model, and there are nvis of them (the window length) in each case.

```
max_range = 0.0 ;
for (icase=0 ; icase<n_cases ; icase++) {
  inptr = ... ;                        // Point to this case in the training set
  sum = xmin= xmax = 0.0 ;             // Start the undifferenced series at zero
  for (icol=0 ; icol<nvis ; icol++) {  // Traverse the window, the series length
    sum += inptr[icol] ;               // Integrate to undifference
    if (sum < xmin)
      xmin = sum ;                     // Minimum of undifferenced series
    if (sum > xmax)
      xmax = sum ;                     // And its maximum across the window
    }
  if (xmax - xmin > max_range)         // Max range across all training cases
    max_range = xmax - xmin ;
  }
```

Now suppose we have generated a sample, and its nvis (the window length) values are in the array genptr. Keep in mind that to train the RBM we have almost certainly rescaled the actual predictors to have an RBM-friendly range of 0–1. Naturally, the generated sample will be in this same 0–1 rescaled domain. But the xmin and xmax computed in the prior code fragment are from the training data *prior* to being rescaled 0–1. So, don't neglect to un-rescale the elements of genptr back to the training data's domain before executing the following code; do this the reverse of how you rescaled the data to 0–1. Then we trivially compute the minimum and maximum of the undifferenced generated series as shown here:

```
sum = xmin = xmax = 0.0 ;             // Start the generated series at 0.0
for (icol=0 ; icol<nvis ; icol++) {   // image_cols = nvis+1
  x = genptr[icol] ;                  // Value for this window position
  sum += x ;                          // Undifference to get series
```

```
    if (sum < xmin)                    // Keep track of its min and max
       xmin = sum ;
    if (sum > xmax)
       xmax = sum ;
    }
```

The code just shown walks across the window, converting the generated differences back into the series domain by summing (undifferencing). We now know the range of this sample series.

As discussed earlier, we want the display of the generated samples to be scaled according to the natural scale of the original data. Thus, we want the xmin-to-xmax range of the generated series just computed to map to a display row range of image_rows/max_range, where image_rows is the number of rows in our visual display, and max_range is the maximum range of any series in the training set. This way, a generated series that happens to exactly match the maximum in the training set will also occupy the maximum range of the display.

We could just place the display at the bottom of the display rectangle, but I find it to be more visually appealing to vertically center the series in the display box. To do this, we need to compute how many rows this sample will occupy and split the difference. This leads us to three lines of code for scaling and centering, as shown here:

```
scale = image_rows / max_range ;    // Factor maps series to display
ioff = (int) (scale * (xmax - xmin)) ;  // Number of rows covered by plot
ioff = (image_rows - ioff) / 2 ;    // Split the difference to center plot in box
```

The final step is to map the series to the image display rectangle. The code to do this is shown next. We march along columns, left to right, summing the generated differences as we go and plotting the series in the display rectangle.

```
sum = 0.0 ;                                     // Series starts at zero
for (icol=0 ; icol<image_cols ; icol++) {       // Columns in display = nvis+1
   for (irow=0 ; irow<image_rows ; irow++)      // Init this column to white
      image_ptr[irow*image_cols+icol] = (unsigned char) 255 ; // White is 255
   irow = (int) (scale * (sum - xmin)) + ioff ;   // Row corresponding to this point
   if (irow >= image_rows)                      // Should be extremely rare
      irow = image_rows - 1 ;
   irow = image_rows - 1 - irow ;               // Mirror vertically; increase is up
```

```
      image_ptr[irow*image_cols+icol] = (unsigned char) 0 ; // Black
      if (icol < nvis) {                  // Must not do this past end of data
         x = genptr[icol] ;               // Difference for this column
         sum += x ;                       // Undifference to get series
         }
      } // For all columns
```

The code just shown sets the entire display box to white, except for a single black entry in each column, with the row of this black entry being defined by the value of the series at that column.

Note that there will be one more image column than generated differences because the first column corresponds to a series value of zero, the initial value of the undifferenced series. For this reason we need the if(icol<nvis) check at the end of the loop, where nvis is the window length (number of generated differences). This check will prevent access of nonexistent data for the last column.

Also note that we want to follow the standard convention that a graph display is such that larger values of the series are higher on the display. So, we mirror vertically to avoid displaying the series upside-down!

Figure 2-3 is an example of some of the generative patterns that an RBM learned from a 20-day window in the S&P 100 Index OEX. The predictors are the difference of logs.

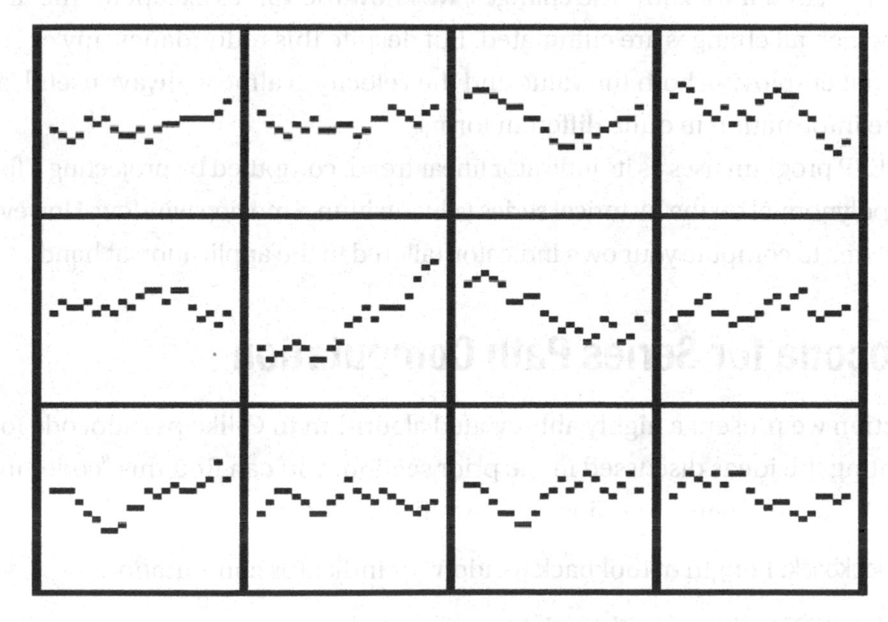

***Figure 2-3.** Patterns from a 20-day window in OEX*

Path of a Function

It's common to apply a function to recent historical values of a time series to compute variables suitable for input to a model. For example, in financial market prediction, we might measure recent trend and volatility. Perhaps each day we would look at the trend of the market over the most recent 10 days. We might also measure volatility over this same time period. Then, these two variables would serve as inputs to a prediction model (perhaps with other families of variables as well).

But if we have a model with relatively little danger of overfitting, such as a deep belief net, we could add an immense amount of information by looking at how a function is *evolving* recently. For example, we might define the function as the trend over a 10-day period. Then we could evaluate this function for the 10-day period ending today, and the 10-day period ending yesterday, and the 10-day period ending the day before, and so on. This collection of 10-day values shows not just what the trend is at this moment in time but also how this trend has evolved in recent days. The trend may be steadily increasing, or it may have just reversed, or whatever.

There are two basic ways of presenting this information. We could look at the actual values of the function over this set of historical measurement periods, or we could examine the day-to-day change (*velocity*) in the function. Obviously, these two approaches are almost perfectly redundant: if we know the values, we know all changes except the first, and if we know the changes, we know the values except for the first, the one from which all changes are cumulated. But despite this redundancy, my experience has been that employing both the value and the velocity is almost always useful, as they present the information in quite different forms.

The DEEP program uses as its indicator linear trend, computed by projecting a first-order Legendre polynomial on the historical series values within a moving window. However, you should feel free to compute your own indicator, tailored to the application at hand.

Pseudocode for Series Path Computation

In this section we present a highly abbreviated algorithm in C-like pseudocode for implementing the ideas discussed in the prior section. You can find this "code" in the file SERIES.TXT. The following quantities are specified by the user:

`lookback`: Length of lookback window for indicator computation

`length`: Number of basic variables generated

also_velocity: 1 or 0; if 1, also generate velocity variables

shift: Number of observations to shift window for each generated case

nature: One of the following four values that control the computation

RAW: Use the actual observations

RAWLOG: Use the log of the actual observations

DIFF: Use the difference of consecutive observations for targets

DIFFLOG: Use the difference of the logs of consecutive observations for targets

The presence or absence of the LOG specification impacts both predictors (log, if specified, is taken before computing the trend) and the target. But the presence or absence of the DIFF specification affects only target computation. This is because in most applications, it would make little or no sense to difference the source series before computing the indicator. In those rare situations in which such pre-differencing is needed, the user should perform the differencing prior to this computation.

We generate length basis predictors. If the user requests that velocity variables also be generated, the number of predictors is doubled. Allocate three vectors for computation.

```
// Find number of predictors generated
npred = length ;
if (also_velocity)
   npred *= 2 ;

window = [Allocate memory for length+also_velocity observations] ;
buffer = [Allocate memory for lookback+1 observations] ;
coefs = [Allocate memory for lookback observations] ;
```

Compute and save the first-order Legendre polynomial coefficients for defining trend. We will compute the trend of a series as the projection of this vector onto the series.

```
sum = 0.0 ;
for (i=0 ; i<lookback ; i++) {
   coefs[i] = 2.0 * i / (lookback - 1.0) - 1.0 ;
   sum += coefs[i] * coefs[i] ;
   }
```

```
sum = sqrt ( sum ) ;
for (i=0 ; i<lookback ; i++)
  coefs[i] /= sum ;
```

Initialize and start the main processing loop. We have to do double buffering because the first buffer cumulates series values for computing the indicator, and the second buffer cumulates indicator and target values for outputting to the database.

```
shift_count = shift-1 ;   // Counter for window shifting
n_in_buffer = 0 ;         // Must fill buffer before we start computing indicators
n_in_window = 0 ;         // Must fill window before we start generating database cases
n_cases = 0 ;             // This will count cases

Initialize 'Most recent value of source series' to the first observation

for (;;) {
  x = [Most recent value of source series] ;
```

If the buffer is full, shift down and put the new one in. We always keep one sample ahead in the buffer for the target.

```
  if (n_in_buffer == lookback+1) {           // Buffer is full?
    for (i=1 ; i<n_in_buffer ; i++)          // Yes, so shift old values down
      buffer[i-1] = buffer[i] ;
    buffer[n_in_buffer-1] = x ;              // Insert next value at end of buffer
    }

  else
    buffer[n_in_buffer++] = x ;              // Keep filling the buffer
```

If the buffer is not yet full, do nothing more. Just loop back to get the next element from the source series. But if it is full, we can compute the trend indicator (or whatever indicator is desired). We need to keep one extra element in the buffer, beyond the lookback needed to compute the indicator. This one extra value is for computing the target. The trend is computed by finding the dot product of the Legendre polynomial with the values in the historical window, possibly modified by first taking logs if the user wants.

```
if (n_in_buffer < lookback+1)  // Do nothing if it's not full yet
  continue ;

x = 0.0 ;
if (nature == RAWLOG || nature == DIFFLOG ) {
  for (i=0 ; i<lookback ; i++) {
    if (buffer[i] > 0.0)
      x += coefs[i] * log ( buffer[i] ) ;
    else
      x -= 1.e60 ;
    }
  }

else {
  for (i=0 ; i<lookback ; i++)
    x += coefs[i] * buffer[i] ;
  }
```

Now we handle the second level of buffering. If the indicator output window is full, we move down the prior values and insert this new indicator value into the last slot. Otherwise, we keep filling the window. If we are also computing velocity, we need one extra slot filled.

```
if (n_in_window == length+also_velocity) {
  for (i=1 ; i<n_in_window ; i++)
    window[i-1] = window[i] ;
  window[n_in_window-1] = x ;
  }

else
  window[n_in_window++] = x ;       // Keep filling the window
```

If the output window is not yet filled, we just loop back and keep filling it. But if it is full, we are ready to proceed. Keep track of whether we have shifted the requested number of slots.

```
if (n_in_window < length+also_velocity) // Do nothing if it's not full yet
  continue ;

if (++shift_count < shift)
  continue ;
shift_count = 0 ;
```

We can now output the predictors. If the user requested velocity, we output variables in pairs, the trend first, followed by the velocity (change in trend).

```
if (also_velocity) { // If true, npred = 2 * length; else npred = length
  for (i=1 ; i<=length ; i++) {
    record[2*i-2] = window[i] ;                    // Indicator
    record[2*i-1] = window[i] - window[i-1] ;      // Velocity
    }
  }

else {
  for (i=0 ; i<npred ; i++)
    record[i] = window[i] ;
  }
```

The last major step is to compute and output the target. This comes from the raw data buffer. There are four possible ways the target is defined: with/without logs and with/without differencing.

```
if (nature == RAW)
  record[npred] = buffer[lookback] ;

else if (nature == RAWLOG) {
  if (buffer[lookback] > 0.0)
    record[npred] = log ( buffer[lookback] ) ;
  else
    record[npred] = -1.e60 ;
  }

else if (nature == DIFF)
  record[npred] = buffer[lookback] - buffer[lookback-1] ;
```

```
else if (nature == DIFFLOG) {
  if (buffer[lookback] > 0.0)
    tempa = log ( buffer[lookback] ) ;
  else
    tempa = -1.e60 ;
  if (buffer[lookback-1] > 0.0)
    tempb = log ( buffer[lookback-1] ) ;
  else
    tempb = -1.e60 ;
  record[npred] = tempa - tempb ;
  }
```

All that remains is final cleanup for this case. Copy record to the database if we did not put it there as part of computation. Remember that the target is the last item in the record, after all predictors. Advance to the next observation in the source series.

```
Output record to database
Advance source series to next observation
++n_cases ;   // Count cases generated
} // For all cases
```

Example of Path Series Creation

This section presents an example of how we can create path-based predictors and targets from a series. The source series is prices of the S&P 100 Index OEX. Because this is an equity price series, our predictors and target are based on log prices. The target is the difference of log prices, while the trend and velocity-of-trend predictors do not involve differencing of the prices. The DEEP.LOG file will be discussed in greater detail in Chapter 6, so here we will focus exclusively on the content of the file as it relates to generating these predictor and target variables for model building. The log file records the user's choices.

Most of the user's choices shown in the Figure 2-4 log file excerpt are clear, as they have already been discussed. But there are a few items to note:

- It correctly states that the log of the series is differenced to produce the target, but only logs are used for the predictors, with no differencing of the raw prices.

- The window length is 20, but because the user selected that velocity also be computed, there are 40 predictors: 20 trends and 20 trend differences.
- Only the target is trimmed. The effect of combining many prices to compute a trend quashes all but the most severe outliers, so there is little point in trimming raw prices before computing trend. If the user anticipates severe outliers, the data should be trimmed before being read by the DEEP program.
- Variables occur in value/velocity pairs. For example, the first two predictors are Value_19 and Veloc_19. These are the trend whose computation window ends 19 bars prior to the current bar, and that value minus the trend with the computation window placed one bar earlier, respectively.
- The predictive target (Lead_1) and the classification targets (Lead_Pos and Lead_Neg) are exactly as in a simple series.

```
Reading series file D:\DEEP\TEST\OEX_TRAIN.TXT for 'path' variables
  The log of the series is differenced to create the target;
    log of series for source
  Data is extracted from column 2
  Lookback for trend is 40
  Window length is 20
    (With velocity, this gives 40 predictors)
  Shift (spacing of adjacent cases in series) is 1
  Tails are trimmed by 5.000 percent
  Targets are multiplied by 1.000
  Above/below median class variables are generated and model is classifier
  No header record is skipped

Path series target trimming is as follows...
       Variable       Old min      New min      New max      Old max

         Lead_1      -0.23540     -0.01745      0.01722      0.10151

Path series defined for training; median split = 0.00050

Basic statistics after any trimming...
       Variable        Mean          StdDev          Min           Max
       Value_19       0.02302       0.12921       -0.71782       0.39170
       Veloc_19      -0.00007       0.01032       -0.08359       0.05897
       Value_18       0.02295       0.12915       -0.71782       0.39170
       Veloc_18      -0.00007       0.01032       -0.08359       0.05897
       ...
        Value_1       0.02228       0.12864       -0.71782       0.39170
        Veloc_1      -0.00003       0.01031       -0.08359       0.05897
        Value_0       0.02225       0.12861       -0.71782       0.39170
        Veloc_0      -0.00003       0.01030       -0.08359       0.05897
         Lead_1       0.00034       0.00878       -0.01745       0.01722
       Lead_Pos       0.49993       0.50000        0.00000       1.00000
       Lead_Neg       0.50007       0.50000        0.00000       1.00000

Sucessfully read 7613 cases with 43 variables
```

Figure 2-4. *DEEP.LOG record of path series generation*

Fourier Coefficients in a Moving Window

When a time series contains a significant quantity of one or more periodic components, the Fourier coefficients generated from a (usually) short window ending at the current time can carry predictive information.

In some situations, the magnitude of the coefficients alone carries the useful information; the phase is irrelevant. For example, perhaps the series represents the acoustic output of a mechanical device. Normally, the high-frequency components are minimal. But if failure is immanent, weakened gears may generate high-frequency artifacts. It is only the presence or absence of such artifacts that is important.

But more often, the phase is a critical piece of information. For example, if a prominent periodic component is at a phase such that as of the end of the window the component is decreasing, we would expect that at the upcoming future observation the down-trend would continue. The converse is also true, of course. In such a situation, lack of phase information would cripple the application.

For this reason, my recommendation is to always (or at least by default) supply both the real and imaginary components to a model. Do not be tempted to instead supply the magnitude and phase to the model, even though they convey exactly the same information. Although the magnitude is easily processed by most models, the discontinuity inherent in a phase measurement makes phase nearly useless in all but a very few highly specialized applications.

In nearly all applications it is crucial to taper the raw data within the window in order to alleviate wraparound effects and side lobes. My own favorite is the Welch window given by Equation 2-1. In this equation, n is the number of data points in the window. This function is one at the center, and it drops to nearly zero at the ends.

$$m_i = 1 - \left(\frac{i - 0.5(n-1)}{0.5(n+1)} \right)^2, \quad i = 0, \ldots, n-1 \tag{2-1}$$

To normalize the Fourier coefficients to compensate for the downward bias produced by the window and also to make values independent of the window length, we usually modify the window as shown in Equation 2-2.

$$m'_i = \frac{m_i}{\sqrt{n \sum_{i=0}^{n-1} m_i^2}} \tag{2-2}$$

If the data is naturally centered around zero, this is all we need to worry about. But if the data has a significant offset from zero, tapering it down to zero at the ends will introduce spurious low-frequency components. For example, consider a fairly flat series with a large positive mean. If we use a Welch window to taper the ends of this series toward zero, we will end up with something that looks like Figure 2-5. Clearly, this naive taper is massively disruptive. In this case, we must compute the window-weighted mean of the raw series as shown in Equation 2-3. This quantity is subtracted from each raw data point before application of the Welch window.

$$\overline{x} = \frac{\sum_{i=0}^{n-1} m_i \, x_i}{\sum_{i=0}^{n-1} m_i} \tag{2-3}$$

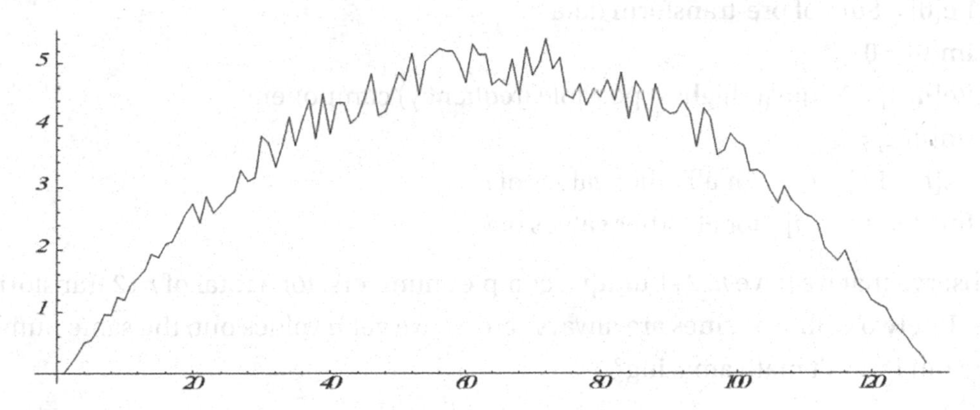

***Figure 2-5.** Positive-mean data after Welch window*

A code fragment for performing these operations is shown next. In this code, there are n data points in the array xr.

```
wsum = dsum = wsq = 0.0 ;
for (i=0 ; i<n ; i++) {
   win = (i - 0.5 * (n-1)) / (0.5 * (n+1)) ;
   win = 1.0 - win * win ;              // Welch data window: Equation 2-1
   wsum += win ;                        // Needed for Equation 2-3
   dsum += win * xr[i] ;                // Needed for Equation 2-3
   wsq += win * win ;                   // Needed for Equation 2-2
   }

wsq = 1.0 / sqrt ( n * wsq ) ;          // Equation 2-2
dsum /= wsum ;                          // Equation 2-3

for (i=0 ; i<n ; i++) {
   win = (i - 0.5 * (n-1)) / (0.5 * (n+1)) ;
   win = 1.0 - win * win ;              // Welch data window: Equation 2-1
   win *= wsq ;                         // Equation 2-2
   xr[i] = win * (xr[i] - dsum) ;       // Window after centering per Equation 2-3
   }
```

A complete Fourier transform of a real series will produce n complex numbers, which will have certain symmetries and redundancies that must be taken into account when using them as inputs to a model. They can be summarized as follows:

If n is even
  Re[0] = Sum of pre-transform data
  Im[0] = 0
  Re[n/2] = Nyquist (highest possible frequency) component
  Im[n/2] = 0
  Re[i] = Re[n-i] for all other values of i
  Im[i] = -Im[n-i] for all other values of i

Observe that we have n/2+1 unique complex numbers, for a total of n+2 transformed values. But two of those values are always zero, so we get n values out, the same number that we put in. Isn't math amazing?

If n is odd
  Re[0] = sum of pre-transform data
  Im[0] = 0
  Re[i] = Re[n-i] for all other values of i
  Im[i] = -Im[n-i] for all other values of i
  Re[n/2] and Im[n/2] are the almost-Nyquist components
   (n/2 is half of n, dropping the fraction.)

Again, observe that we get n unique values out, the same number that we put in.

Remember that if we center using Equation 2-3, the pre-transform sum is zero, so Re[0] = 0.

These facts are important because we should not supply redundant information to the model, nor should we use as inputs any values that we know in advance will always be zero!

The free source file collection that can be downloaded from my web site includes a set of fast Fourier transform routines that will handle fully complex series of any length. Complete instructions for using these routines are included in the comments section at the beginning of MRFFT.CPP. These routines also include the ability to efficiently transform even-length real-valued series by packing the real input series into the real and imaginary inputs to the complex transform. Documentation for this somewhat more efficient technique is included in the same source file.

Pseudocode for Fourier Series Processing

In this section, we present a highly abbreviated algorithm in C-like pseudocode for implementing the generation of predictors and targets by means of a Fourier transform of a moving window. You can find this "code" in the file SERIES.TXT.

We begin with a rigorous presentation about which transformed values go where in the database, as well as which of these variables are assigned to the default predictor list for the user's convenience. Two general rules govern these assignments:

- Any value that is known from theory to always be zero is omitted from the database.
- The default predictor set is entirely arranged as real/imaginary pairs.

These general rules lead to the following specific rules and side effects:

- Im[0] never goes into the database because it is always zero.
- Re[0] goes into the database if and only if the user requests no centering. Centering zeros it. It never goes into the default predictor set because it does not have an imaginary partner.
- Re[n/2] always goes into the database. If n is even this is the Nyquist frequency. If n is odd, this is the "almost-Nyquist" frequency.
- Im[n/2] goes into the database if and only if n is odd. If n is even it is always zero.
- If n is odd, Re[n/2] and Im[n/2] both go into the default predictor set. If n is even, neither goes into the default predictor set, because this would break pairing (Im[n/2] is not in the database). If the user wants Re[n/2] as predictor, it must be added manually.

Here is a more rigorous statement of the algorithm by which Fourier variables go into the database and default predictor set. In this algorithm, n_vars is the number of variables in the database, n_pred is the number of variables in the default predictor set, center is true if the user requests centering, and length is n, the number of observations in the data window.

```
n_vars = n_pred = 0 ;

if (! center) {
  Put name "Offset" in database name array ;
  ++n_vars ;
  }

for (i=1 ; i<length/2 ; i++) {
  Put name "Real_i" in dataset name array ;
  ++n_vars ;
  Put name "Real_i" in default predictor list
  ++n_pred ;
  Put name "Imag_i" in dataset name array ;
  ++n_vars ;
  Put name "Imag_i" in default predictor list
  ++n_pred ;
  }

Put name "Real_n/2" in database name array ;
++n_vars ;
if (length % 2) { // Is length odd?
  Put name "Real_n/2" in default predictor list
  ++n_pred ;
  Put name "Imag_n/2" in database name array ;
  ++n_vars ;
  Put name "Imag_n/2" in default predictor list
  ++n_pred ;
  }
```

Here is the workhorse routine that transforms a window and returns a set of complex-domain Fourier coefficients. The Welch window already described is included in this listing, despite being redundant. It's useful to see it in context. The variables are put into the out array in the same order as the database variable names were inserted in the algorithm just shown.

```
void do_fft (
  int n ,           // Length of source data window; number of cases in 'in'
  int center ,      // If nonzero, center the data before transforming (recommended)
  double *in ,      // Source data, n long, preserved
  double *out ,     // Output of transform
  double *work ,  // Work array 2*n long
  FFT *fft
  )
{
  int i, k ;
  double *xr, *xi, win, wsum, dsum, wsq ;

  xr = work ;
  xi = xr + n ;

/*
  It would be slightly more efficient to use the half-length FFT
  method. But the difference is tiny and not worth the bother of
  dealing with possibly odd length series.
*/

  for (i=0 ; i<n ; i++) {
    xr[i] = in[i] ;
    xi[i] = 0.0 ;
    }

  wsum = dsum = wsq = 0.0 ;
  for (i=0 ; i<n ; i++) {
    win = (i - 0.5 * (n-1)) / (0.5 * (n+1)) ;
    win = 1.0 - win * win ; // Welch data window
    wsum += win ;
    dsum += win * xr[i] ;
    wsq += win * win ;
    }

  if (center)
    dsum /= wsum ;                  // Weighted mean
  else
    dsum = 0.0 ;
```

```
   wsq = 1.0 / sqrt ( n * wsq ) ;          // Compensate for reduced power

   for (i=0 ; i<n ; i++) {
      win = (i - 0.5 * (n-1)) / (0.5 * (n+1)) ;
      win = 1.0 - win * win ;              // Welch data window
      win *= wsq ;                         // Compensate for reduced power
      xr[i] = win * (xr[i] - dsum) ;       // Window after centering
      }

   fft->cpx ( xr , xi , 1 ) ;              // Transform to frequency domain

   k = 0 ;                                 // Will index entries in database

   if (! center)
      out[k++] = xr[0] ;                   // Re[0]

   for (i=1 ; i<n/2 ; i++) {               // Central frequencies
      out[k++] = xr[i] ;
      out[k++] = xi[i] ;
      }

   out[k++] = xr[n/2] ;                    // Re[n/2]
   if (n % 2)                              // If n is odd
      out[k++] = xi[n/2] ;                 // Im[n/2]
}
```

Here is the pseudocode showing the algorithm that oversees collecting observations from the moving window, transforming, and outputting the Fourier variables and the targets to the database.

The following quantities are specified by the user:

length: Number of historical predictors desired

center: If nonzero, center the data before transforming (usually good!)

shift: Number of observations to shift window for each generated case nature: One of the following four values:

RAW: Use the actual observations

RAWLOG: Use the log of the actual observations

DIFF: Use the difference of consecutive observations

DIFFLOG: Use the difference of the logs of consecutive observations

The required number of cases in the source series moving window is length, plus one extra for the target (which is the next observation after the predictors). Allocate memory and create the FFT object. Identify which variables will go into the database and which into the default predictor list, as already described. Initialize for the main processing loop.

```
window = [Allocate memory for length+1 observations] ;
fft_work = [Allocate memory for 2*length observations] ) ;
fft = new FFT ( length , 1 , 1 ) ;     // Code and comments in MRFFT.CPP

[Set names for the database and default predictor list variables, as described earlier.]

n_in_window = 0 ;        // Must fill window before we start generating database cases
shift_count = shift-1 ;  // Counter for window shifting
n_cases = 0 ;            // Will count generated cases
```

The main loop is here. We begin with the first data point, filling the data window and then transforming and moving the window forward in time. If the user requested taking logs before the transform, do it.

```
Initialize 'Most recent value of source series' to the first observation

for all cases {

  x = [Most recent value of source series] ;

  if (nature == RAWLOG || nature == DIFFLOG) {
    if (x > 0.0)
      x = log ( x ) ;
    else
      x = -1.e60 ;
    }
```

```
if (n_in_window == length+1) { // If window is full, shift down and put new case in
  for (i=1 ; i<n_in_window ; i++)
    window[i-1] = window[i] ;
  window[n_in_window-1] = x ;
  }

else
  window[n_in_window++] = x ; // Keep filling the window
if (n_in_window < length+1) { // Do nothing if it's not full yet
  Advance source series to next observation
  continue ;
   }
```

When we get here, the window is full, so we are ready to compute the variables for this window. But check how much the user wants to shift the window between variables for the database.

```
if (++shift_count < shift) {
  Advance source series to next observation
  continue ;
  }

shift_count = 0 ;
```

Compute the variables. Recall that do_fft() puts the variables into the output array (record here) in the order shown earlier. As usual, the target goes into the record array after the n_vars Fourier variables. The target is either the source observation or the difference from the prior observation. If the user requested taking logs, this has already been done for the data in window.

```
do_fft ( length , center , window , record , fft_work , fft ) ;

if (nature == RAW || nature == RAWLOG)
  record[n_vars] = window[n_in_window-1] ;   // Target
else if (nature == DIFF || nature == DIFFLOG)
  record[n_vars] = window[n_in_window-1] - window[n_in_window-2] ;
```

```
Output record to database
Advance source series to next observation
++n_cases ;   // Count cases generated
} // For all cases
```

Example of Fourier Series Generation

Perhaps the prior sections have thoroughly pounded into your head the rules by which Fourier variables are written to the database and set as default predictors. But at the risk of being pedantic, this section will present the database and default predictor set variables in four scenarios. In each case, the first table will list the variables in the database, and the second table will list the default predictors. You are strongly encouraged to observe, in each case, the presence or absence of the offset and real/imaginary Nyquist terms in accordance with our rules.

Length=10 with No Centering

Variable	Mean	StdDev	Min	Max
Offset	5.41045	0.70241	3.75781	6.37693
Real_1	-1.16488	0.15121	-1.38084	-0.80157
Imag_1	0.37801	0.04979	0.24152	0.46762
Real_2	-0.23301	0.03030	-0.28438	-0.16026
Imag_2	0.16916	0.02217	0.11122	0.21247
Real_3	-0.06496	0.00858	-0.08473	-0.02893
Imag_3	0.08930	0.01175	0.05380	0.11447
Real_4	-0.01303	0.00221	-0.02516	0.00764
Imag_4	0.03994	0.00539	0.01513	0.05331
Real_5	-0.00004	0.00192	-0.01483	0.01579
Lead_1	0.00037	0.00882	-0.01745	0.01736
Lead_Pos	0.52812	0.49921	0.00000	1.00000
Lead_Neg	0.47188	0.49921	0.00000	1.00000

```
Weights for final (output) layer
          0.002624  Real_1
          0.001881  Imag_1
          0.000831  Real_2
```

```
 0.002511  Imag_2
 0.001193  Real_3
 0.000572  Imag_3
 0.000081  Real_4
-0.000471  Imag_4
 0.000366  CONSTANT
```

Length=10 with Centering

```
Variable       Mean        StdDev        Min         Max

 Real_1     -0.00014      0.00374    -0.02235     0.04431
 Imag_1     -0.00044      0.00601    -0.02921     0.07200
 Real_2     -0.00006      0.00249    -0.03141     0.03547
 Imag_2     -0.00009      0.00255    -0.02658     0.03967
 Real_3     -0.00005      0.00179    -0.01782     0.02659
 Imag_3     -0.00003      0.00177    -0.02052     0.02024
 Real_4     -0.00004      0.00146    -0.01406     0.01882
 Imag_4     -0.00001      0.00143    -0.01863     0.01886
 Real_5     -0.00004      0.00192    -0.01483     0.01579
 Lead_1      0.00037      0.00882    -0.01745     0.01736
Lead_Pos     0.52812      0.49921     0.00000     1.00000
Lead_Neg     0.47188      0.49921     0.00000     1.00000

Weights for final (output) layer
        -0.000288  Real_1
         0.000402  Imag_1
        -0.000187  Real_2
         0.000471  Imag_2
         0.000122  Real_3
         0.000250  Imag_3
         0.000016  Real_4
        -0.000022  Imag_4
         0.000366  CONSTANT
```

Length=11 with No Centering

```
Variable        Mean      StdDev       Min         Max

 Offset      5.39522     0.70031    3.74946     6.35850
 Real_1     -1.20356     0.15620   -1.42617    -0.82738
 Imag_1      0.35287     0.04662    0.22213     0.43687
 Real_2     -0.25127     0.03267   -0.30672    -0.17402
 Imag_2      0.16135     0.02118    0.10645     0.20109
 Real_3     -0.07795     0.01024   -0.10082    -0.04114
 Imag_3      0.08986     0.01183    0.05637     0.11529
 Real_4     -0.02168     0.00316   -0.03324     0.00452
 Imag_4      0.04736     0.00633    0.02161     0.06412
 Real_5     -0.00218     0.00109   -0.01010     0.01231
 Imag_5      0.01491     0.00244    0.00161     0.02840
 Lead_1      0.00037     0.00882   -0.01745     0.01736
Lead_Pos     0.52819     0.49920    0.00000     1.00000
Lead_Neg     0.47181     0.49920    0.00000     1.00000

Weights for final (output) layer
          0.002203   Real_1
          0.001623   Imag_1
         -0.000057   Real_2
          0.001549   Imag_2
          0.001271   Real_3
          0.000665   Imag_3
          0.000437   Real_4
         -0.000351   Imag_4
          0.000040   Real_5
          0.000159   Imag_5
          0.000366   CONSTANT
```

Length=11 with Centering

Variable	Mean	StdDev	Min	Max
Real_1	-0.00014	0.00379	-0.02195	0.04193
Imag_1	-0.00049	0.00627	-0.02898	0.07294
Real_2	-0.00006	0.00256	-0.03256	0.03085
Imag_2	-0.00009	0.00263	-0.02453	0.04424
Real_3	-0.00005	0.00181	-0.01829	0.02531
Imag_3	-0.00004	0.00180	-0.01798	0.02554
Real_4	-0.00004	0.00149	-0.01302	0.02287
Imag_4	-0.00002	0.00146	-0.01906	0.01681
Real_5	-0.00004	0.00106	-0.00830	0.01412
Imag_5	-0.00001	0.00148	-0.01193	0.01193
Lead_1	0.00037	0.00882	-0.01745	0.01736
Lead_Pos	0.52819	0.49920	0.00000	1.00000
Lead_Neg	0.47181	0.49920	0.00000	1.00000

```
Weights for final (output) layer
        -0.000497  Real_1
         0.000477  Imag_1
        -0.000425  Real_2
         0.000463  Imag_2
        -0.000009  Real_3
         0.000362  Imag_3
         0.000112  Real_4
         0.000109  Imag_4
         0.000028  Real_5
         0.000170  Imag_5
         0.000366  CONSTANT
```

One more crucial item to observe in the prior tables concerns the effect of centering versus not centering. As is common, the source data is strongly offset from zero. When you apply the Welch window, if the data is not centered, you suffer severe spurious low-frequency contamination. This is why centering is *required* unless the data is naturally centered.

Morlet Wavelets

We often have a process that includes periodic behavior. There may be a physical structure involved (such as a vibrating or rotating component) that produces this effect, or "ringing" may result from an unexpected disruption. A classic example is in financial markets, when a profound world event causes a rapid price change that overshoots the "fair" value, followed by an excessive correction in the opposite direction, and perhaps a few more diminishing oscillations before the price settles.

In such applications, we may want to answer one or more of the following questions:

- When did the periodic phenomenon start?
- When did it stop?
- Where in its cycle is it right now (peak, trough, rising, falling)

Morlet wavelets provide an effective way to answer these questions. This subject is much too large to handle in depth in this text. But we'll summarize a few key points and present the core mathematics and source code snippets for computing them. Here are the salient points:

- Morlet wavelets are robust against shifting in time. Regardless of when the phenomenon appears, its signature will remain consistent. This is very much *not* the case for orthogonal wavelets such as the famous Daubechies family.
- In an ideal world we would often like to precisely define the periodic behavior of the phenomenon and also precisely pinpoint its beginning and end. Unfortunately, the *Heisenberg Uncertainty Principle* prevents us from doing both simultaneously. We can precisely define the period but only vaguely locate its beginning and end, or we can pinpoint the time extent of the phenomenon but only if we are loose in its periodicity. This principle rigorously specifies a limit to how well we can do both. The Morlet wavelet is the only wavelet that achieves this limit (or comes very, very close).

The time-domain representation of a Morlet wavelet is given by Equation 2-4, in which w determines the center frequency and s is a scale factor. A typical Morlet wavelet is shown in Figures 2-6 (the real part) and 2-7 (the imaginary part). To compute a Morlet wavelet transform, we could find the dot product of the time-domain representation with the series being processed. However, computing Morlet wavelets is more straightforward in the frequency domain, which we will pursue shortly.

$$h(t) = \pi^{-\frac{1}{4}} \left(e^{2\pi i w t} - e^{-(\pi w s)^2} \right) e^{-t^2/s^2} \tag{2-4}$$

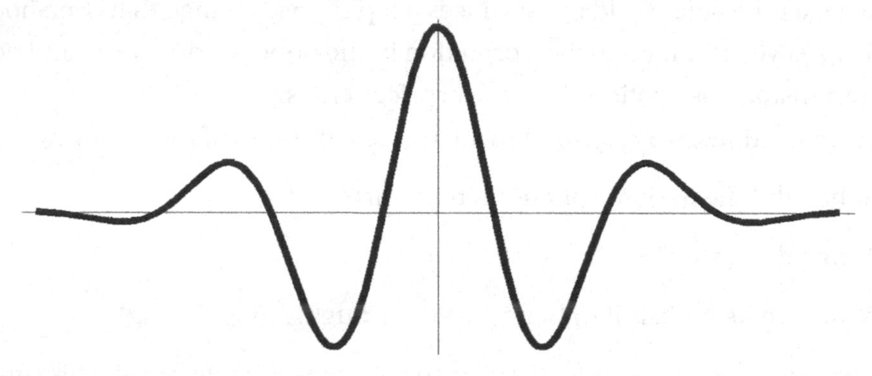

Figure 2-6. *Real component of Morlet wavelet*

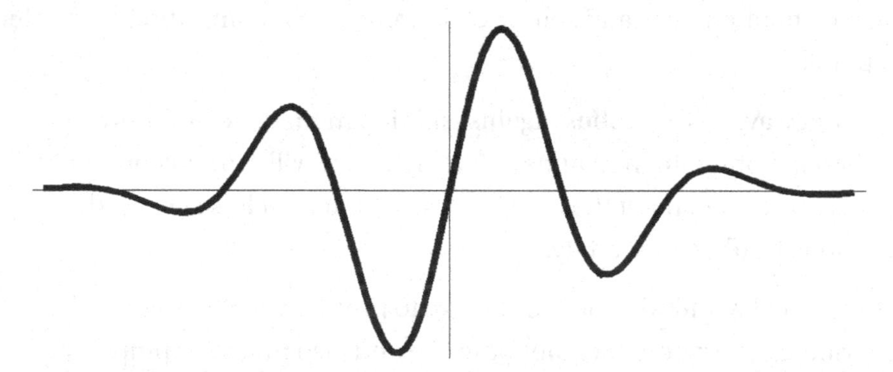

Figure 2-7. *Imaginary component of Morlet wavelet*

An accurate estimate of a Morlet wavelet transform can be easily obtained by finding the Fourier transform of the source series, applying the frequency-domain weighting function (which, except for a constant multiplier, is the Fourier transform of Equation 2-4), and then transforming back to the time domain. For the real component, the frequency-domain function is given by Equation 2-5, and the imaginary component is shown in Equation 2-6. In these equations, *f* is the frequency for which a transform weight is being computed, and it ranges from 0 to 0.5 (the Nyquist frequency). For frequencies between 0.5 and 1.0, which will be encountered if one takes the straightforward approach of a full complex transform, the real part is symmetric around 0.5, and the imaginary part is anti-symmetric.

The center frequency of the filter is *w*, and it has the same range. The scale factor *r* controls the width of the filter, with smaller values creating a narrower response. The time-domain scale factor *s* in Equation 2-4 is related to the frequency-domain scale factor *r* by the relationship $sr=1/\pi$.

$$H_{Re}(f)=e^{-\left(\frac{f-w}{r}\right)^2}+e^{-\left(\frac{f+w}{r}\right)^2}-2e^{-\left(\frac{f^2-w^2}{r^2}\right)} \tag{2-5}$$

$$H_{Im}(f)=i*\left[e^{-\left(\frac{f-w}{r}\right)^2}-e^{-\left(\frac{f+w}{r}\right)^2}\right] \tag{2-6}$$

The real and imaginary parts have complementary interpretations. Look again at Figures 2-6 and 2-7 and imagine forming the dot product of each of these with a time series having at least roughly their periodicity. When the real component is precisely aligned with the wave, its central peak at a series peak, the dot product will be at its maximum. When it is aligned peak-to-trough, it will be at its minimum, a large negative value. When it is shifted one-quarter period, it will be zero. So, the real component reveals the *position* we are at within the period.

When the center (zero crossing) of the imaginary component is aligned at a series peak or trough, the dot product will be zero because of symmetry. When the center is aligned with a zero crossing of the series, the dot product will have a large magnitude. So, the imaginary part tells us the *velocity*, the rate at which the periodic component is changing.

Period, Width, and Lag

The terms shown in Equations 2-5 and 2-6 are needed for computation, but they are a bit too theoretical for most users, who generally think in terms of sample points, not frequencies ranging from 0 to 0.5. Recall that frequency is the reciprocal of period. So, if the user is interested in a phenomenon that repeats with a period of, say, ten samples, then *w* in the previous equations would be 0.1.

It is also easiest to think about the width of a filter in terms of sample points in the source series. Theoretically, a Morlet wavelet transform involves an infinite number of sample points. However, in practical terms, the impact of sample points on the computed real and imaginary components of a wavelet transform covers a relatively short time span. This should be apparent from Figures 2-6 and 2-7. The filter has a center point, the sample that has maximum impact on the computed Morlet wavelet, and the impact of sample points diminishes for samples more distant from the center point. The number of sample points on *each side* of the center that go into computation is called the *width* in this text and in the DEEP program. Many standard references call this the *half-width* because it measures roughly half of the total filter extent.

It turns out that we can conveniently define (for the purposes of this text and DEEP) the frequency-domain scale factor r as 0.8 divided by the *width* just discussed. Recall that $sr=1/\pi$. Simple algebraic substitution tells us that when the time-domain expression of the Morlet wavelet given in Equation 2-4 has the time t equal to the *width*, the exponentially decaying multiplier at the right side of that equation will be e to the power of minus (0.8 π) squared, which is about 0.002 and dropping fast, so it's not unreasonable to say that the filter's response is zero by the time we get to *width* past the center! Defining r to be 0.8 divided by *width* is purely heuristic, based on the reasoning just given, but it works well in practice.

The width relative to the period controls the trade-off between locating a periodic event in time versus specificity in frequency for defining the event. Figures 2-6 and 2-7 are based on the width being twice the period, which is a good starting point for experimenting. Observe that the wavelet has just barely completed two periods on each side of the center before essentially vanishing.

If the width is increased, the wavelet becomes more selective in its frequency response. This is good in applications in which there is a clearly defined frequency (period) for the phenomenon, as might be the case in engineering. But the price paid is less ability to precisely determine when the event starts and stops, as the wavelet is further spread out in time. Conversely, for applications in which the specified period is based more on a hunch than fact and typically shows wide variability, making the width smaller would enable the wavelet to detect a wider range of periodic phenomena. Moreover, we would gain greater ability to locate the event in time, which is always nice.

Keep in mind that the wavelet needs samples on both sides of its center. Because we don't know the future, we cannot center the wavelet at the "current" time; doing so would introduce tremendously destructive distortion by effectively decreeing that the values of the series to the right of center are all zero! To be strictly correct, we must center the wavelet *width* samples prior to the current time to have the full complement of samples. In other words, the information we get about the position and velocity of the periodic component will be at a *lag* equal to the width.

This is annoying, as we almost always want our information to be as current as possible. It is somewhat acceptable to cheat a little and lag the wavelet a bit less than the width; the implicitly zero series values that go into the computation may still be far enough from the center that the weights are small. Once again, look at Figures 2-6 and 2-7 to get a rough idea of the rate of dropoff. But please remember that *any* lag less than the width introduces errors. Be judicious.

Code for Morlet Wavelets

First, I should be clear that there are two minor inefficiencies in the code about to be shown. In practice I believe that they are inconsequential, and they greatly simplify both the coding and the reader's understanding of the algorithm. However, to be totally transparent, here they are:

- Even though the signal series in the presumed application is strictly real, a full complex transform is employed. There is a way to pack real values into a complex vector half the length of the full complex representation, do the transform (in half the computer time), and then unpack the data. The routine MRFFT.CPP, available as a free download from my web site, includes this ability and fully documents it in comments. You should have little trouble making this modification if you want. I do not consider it worth the increased complexity, but you are free to disagree.
- My book *Neural, Novel, and Hybrid Algorithms for Time Series Prediction* (1995) describes a relatively complicated method for computing an excellent approximation to a Morlet wavelet transform that produces the real and imaginary parts simultaneously, although a full complex transform is needed. That method is not used here as I have come to the conclusion that its complexity is excessive.

So, the code shown here uses a full complex transform, and it computes the real and imaginary components separately. In an age of incredibly fast computers, any time saved by more efficient algorithms is invisible in any application that I can imagine, and clarity is king.

Note that the software package available for free download from my web site contains MRFFT.CPP and two helper routines that perform full complex fast Fourier transforms. This code contains complete documentation for its use. The code shown here makes use of this module.

The calling parameter list is as follows:

```
void compute_morlet (
  FFT *fft ,            // Class pointer which does the FFT
  int period ,          // Period (1 / center frequency) of desired filter
  int width ,           // Width on each side of center
  int lag ,             // Lag from current for center of filter; ideally equals width
  int lookback ,        // Number of samples in input buffer
  int n ,               // Lookback plus pad, bumped up to nearest power of two
  double *buffer ,      // Input data
  double *realval ,     // Real value returned here
  double *imagval ,     // Imaginary value returned here
  double *xr ,          // Work vector n long
  double *xi ,          // Ditto
  double *yr ,          // Ditto
  double *yi )          // Ditto
```

The first step is to copy the user's input series to a local work area, as the transform routine will replace its input data with the transform. Later operations are simplified by reversing the time order of the series so that the most recent observation appears first in the work vector.

We also pad the end of the series (the oldest data) with an "innocuous" value for as many points as we need to avoid wraparound effects. The mean of the series is commonly used, although if a better value is known, such as the theoretical mean of the series, or zero, then it should be used. This padding is needed only if the user has (probably unwisely) chosen a lag less than the width. Recall that, whether you see it or not (and you don't in the FFT method!), the implicit time-domain filter convolved with the input series wraps around to the other end of the series. You'd better make sure that these wrapped values are as innocent as possible.

For example, suppose the most recent observation is at index 0, the filter width is 10, and the lag is 8. We will be computing the value at index 8, the center of the filter. The implicit time-domain filter will examine the ten later and ten earlier cases. The later cases are valid, but the ten earlier cases are at indices 7, 6, 5, 4, 3, 2, 1, 0, and, um, -1 and

-2. What the heck are at those negative indices? Actually, they are the last two (oldest) samples! So, we had better pad the end of the series with harmless values.

Finally, although this is not strictly necessary, we extend the padding to make the length a power of two. This almost always speeds the transform, sometimes greatly, because a fast Fourier transform is most efficient when all factors are as small as possible. In fact, a "fast" Fourier transform of a series with a prime number of cases is not fast at all! A short code fragment to do this padding and extension, performed before calling compute_morlet(), is shown next. We also allocate the FFT object.

```
lookback = 2 * width + 1 ;          // We'll need this much data
pad = width - lag ;                 // Prevent wraparound.

for (n=2 ; n<MAXPOSNUM/2 ; n*=2) { // Extend n to a power of 2
  if (n >= lookback+pad)
    break ;
  }

fft = new FFT ( n , 1 , 1 ) ;
```

We now return to the compute_morlet() routine. The code to copy the series with time reversal and then pad with the mean is shown next. We also compute a few constants that will be needed.

```
nyquist = n / 2 ;        // The transform and function are symmetric around this
freq = 1.0 / period ;    // Definition
fwidth = 0.8 / width ;   // The heuristic discussed earlier

mean = 0.0 ;
for (i=0 ; i<lookback ; i++) {
  xr[i] = buffer[lookback-1-i] ;
  xi[i] = 0.0 ;
  mean += xr[i] ;
  }

mean /= lookback ;

while (i<n) {
  xr[i] = mean ;
  xi[i++] = 0.0 ;
  }
```

To compute the real part of the Morlet wavelet, we must transform to the frequency domain, multiply by the values given by Equation 2-5, and then transform back to the time domain. The constant factor multiplier is just to normalize the frequency-domain weights to be 1.0 at the peak. This is a somewhat arbitrary choice, although passing the center frequency unmolested seems sensible.

First we process the symmetric part, the values strictly between a frequency of zero and the Nyquist frequency (0.5). We (probably needlessly) add a tiny number to the denominator of the multiplier to ensure against division by zero. It's just another one of my often pointless habits that is impossible to break.

```
fft->cpx ( xr , xi , 1 ) ; // Transform to frequency domain
multiplier = 1.0 / (morlet_coefs ( freq , freq , fwidth , 1 ) + 1.e-140 ) ;

for (i=1 ; i<nyquist ; i++) {     // Do just symmetric part
   f = (double) i / (double) n ; // This frequency
   wt = multiplier * morlet_coefs ( f , freq , fwidth , 1 ) ;  // Equation 2-5
   yr[i] = xr[i] * wt ;
   yi[i] = xi[i] * wt ;
   yr[n-i] = xr[n-i] * wt ;
   yi[n-i] = xi[n-i] * wt ;
   } // For all unique frequencies strictly between zero and Nyquist
```

The Morlet coefficient at f=0 is zero, so set yr[0] and yi[0] to zero. Also, the imaginary Nyquist value in yi[nyquist] is always zero by definition. This is a real transform, so we need to weight the real Nyquist in yr[nyquist].

```
yr[0] = yi[0] = yi[nyquist] = 0.0 ; // Always true
wt = multiplier * morlet_coefs ( 0.5 , freq , fwidth , 1 ) ;
yr[nyquist] = xr[nyquist] * wt ;
```

The last step is to transform back to the time domain and return the lagged value. Recall that we reversed the time order of the series.

```
fft->cpx ( yr , yi , -1 ) ;          // Back to time domain
*realval = yr[lag] / n ;
```

The imaginary part is handled almost identically. However, the function we multiply by is Equation 2-6 times the imaginary *i*, and it is antisymmetric around the Nyquist point, so we need to take this into account for the frequency-domain multiplication.

Also, because of this antisymmetry, the real part of the function crosses at the Nyquist point, so everything is zero at frequencies of zero and 0.5.

```
multiplier = 1.0 / (morlet_coefs ( freq , freq , fwidth , 0 ) + 1.e-140 ) ;

for (i=1 ; i<nyquist ; i++) {           // Do just symmetric part
   f = (double) i / (double) n ;        // This frequency
   wt = multiplier * morlet_coefs ( f , freq , fwidth , 0 ) ;     // Equation 2-6
   yr[i] = -xi[i] * wt ;                // Recall that i * i = -1
   yi[i] = xr[i] * wt ;
   yr[n-i] = xi[n-i] * wt ;             // Again, i * i = -1 but function is anti-symmetric
   yi[n-i] = -xr[n-i] * wt ;
   } // For all unique frequencies strictly between zero and Nyquist

yr[0] = yi[0] = yr[nyquist] = yi[nyquist] = 0.0 ;        // Definition
```

Finally, we transform back to the time domain and return the lagged value. We need to flip the sign to compensate for the fact the we reversed the time order at the start of this routine. The imaginary component is antisymmetric, so the flow of time matters. The real part is symmetric around the center, so time order does not matter.

```
  fft->cpx ( yr , yi , -1 ) ;          // Back to time domain
  *imagval = -yr[lag] / n ;
}
```

We've thus far overlooked morlet_coefs(), the routine that computes the frequency-domain weights. This routine simply evaluates Equations (2-5 and (2-6. It is shown here:

```
static double morlet_coefs ( double freq , double fcent , double fwidth , int is_real )
{
  double x, term1, term2 ;

  x = fabs ( freq - fcent ) / fwidth ;
  term1 = exp ( -x * x ) ;

  x = (freq + fcent) / fwidth ;
  term2 = exp ( -x * x ) ;
```

```
  if (is_real) {
    x = (freq * freq + fcent * fcent) / (fwidth * fwidth) ;
    return term1 + term2 - 2.0 * exp ( -x * x ) ;          // Equation 2-5
    }
  else
    return term1 - term2 ;                                 // Equation 2-6
}
```

We now switch over to pseudocode to demonstrate how a practical program can use the routines just described to compute Morlet wavelet predictors from a time series. This pseudocode, as well as the code shown earlier, can be found in the file SERIES.TXT. The following quantities are specified by the user:

length: Number of historical values of complex transform to compute

shift: Number of observations to shift window for each generated case

period: 1.0/center frequency of wavelet filter

width: Filter half-width

lag: Lag from current value to filter center, ideally equals width

nature: One of the following four values:

- RAW: Use the actual observations
- RAWLOG: Use the log of the actual observations
- DIFF: Use the difference of consecutive observations for the target
- DIFFLOG: Use the difference of the logs of consecutive observations for the target

We must be clear about the meaning of the length parameter, referred to as *Window* in the DEEP manual. This is the length of the window over which Morlet wavelets are computed. We will generate this many complex-valued Morlet wavelet coefficients for each case. For example, suppose we set length=3. Each case generated will contain six numbers. These will be the real and imaginary components for the most recent source series observation, the second-most recent, and the third-most recent. So, the actual number of historical source series observations will greatly exceed the window length; it will go all the way back to the required lookback for the oldest observation in the window.

Because each window component is a complex number, with a real and an imaginary component, the number of predictors per case, npred, is double the window length. The next few lines were briefly discussed earlier in conjunction with the compute_morlet() routine, but a brief review would not hurt. We will almost certainly want to pad the lookback sample points with innocuous values before computing the Morlet transform. There are two reasons for doing this. First, as long as the user's specified lag is equal to the specified width, the filter will not extend into the unknown future to any practical degree. (Recall that there is no reason to lag further back.) However, the user may want to exercise his right to use any shorter lag, even including zero. This violates the meaning of the filter, but it still provides usable (and perhaps excellent) predictors. So, if lag<width, a straight application of the FFT will cause the filter to wrap around to the oldest points, causing serious errors. Hence, we must pad with at least width–lag "neutral" points to ameliorate this problem.

Second, FFTs are by far most efficient when they are dealing with a number of points that is a power of two. Although not strictly necessary, we bump up to a power of two for the sake of speed.

```
npred = 2 * length ;                // Each transform value is a complex number
lookback = 2 * width + 1 ;          // We'll need this much data to compute transform
pad = width - lag ;                 // Prevent wraparound error

for (n=2 ; n<MAXPOSNUM/2 ; n*=2) { // Bump up to a power of two for FFT speed
   if (n >= lookback+pad)
     break ;
   }
```

We now allocate working memory and create the FFT, and we initialize for the main processing loop. The output window is 2*length long because each output is a complex number. The data buffer is one longer than the transform lookback so we can compute the target.

```
window = [Allocate memory for 2*length observations] ;
buffer = [Allocate memory for lookback+1 observations] ;
xr = [Allocate memory for n reals] ;
xi = [Allocate memory for n reals] ;
```

```
yr = [Allocate memory for n reals] ;
yi = [Allocate memory for n reals] ;

fft = new FFT ( n , 1 , 1 ) ;

n_cases = 0 ;              // This will count cases
n_in_window = 0 ;          // Must fill window before we start generating database cases
n_in_buffer = 0 ;          // Must fill buffer before we start computing transform
shift_count = shift-1 ;    // For shifting window
```

The main loop is here. Fetch the most recent source series observation and take its log if the user requests logs. If the data buffer is full, shift down the existing values and put this new one in the end slot. Otherwise, keep filling the buffer. Remember that we keep one more observation in the buffer than is needed for the transform; this lets us compute the target.

```
for all cases {
  x = [Most recent value of source series] ;

  if (nature == RAWLOG || nature == DIFFLOG) {
    if (x > 0.0)
      x = log ( x ) ;
    else
      x = -1.e60 ;
    }

  if (n_in_buffer == lookback+1) {
    for (i=1 ; i<n_in_buffer ; i++)
      buffer[i-1] = buffer[i] ;
    buffer[n_in_buffer-1] = x ;
    }
  else
    buffer[n_in_buffer++] = x ; // Keep filling the buffer
```

If the data buffer is not yet full, we just loop back to keep filling it.

```
  if (n_in_buffer < lookback+1) {   // Do nothing if it's not full yet
    Advance source series to next observation
    continue ;
    }
```

We get here when the data buffer is full, so now we can compute the Morlet transform. If the output window is full, shift down the existing values and put this new complex pair in the end slots. Otherwise, keep filling the output window.

```
compute_morlet ( fft , period , width , lag , lookback , n ,
           buffer , &realval , &imagval , xr , xi , yr , yi ) ;

if (n_in_window == length) {                     // If the output window is full
  for (i=1 ; i<n_in_window ; i++) {              // Shift down existing outputs
    window[2*i-2] = window[2*i] ;
    window[2*i-1] = window[2*i+1] ;
    }
  window[2*n_in_window-2] = realval ;    // Insert this new (complex) output
  window[2*n_in_window-1] = imagval ;
  }

else {
  window[2*n_in_window] = realval ;       // Keep filling the window
  window[2*n_in_window+1] = imagval ;
  ++n_in_window ;
  }

if (n_in_window < length) {   // Do nothing if it's not full yet
  Advance source series to next observation
  continue ;
  }
```

The window is full, so we are ready to output the variables for this window. But check how much the user wants us to shift the window between variables for the database.

```
if (++shift_count < shift)
  continue ;
shift_count = 0 ;

for (i=0 ; i<npred ; i++)
  record[i] = window[i] ;

if (nature == RAW || nature == RAWLOG)
  record[npred] = buffer[lookback] ;
```

```
else if (nature == RAWLOG || nature == DIFFLOG)
   record[npred] = buffer[lookback] - buffer[lookback-1] ;

Output record to database
Advance source series to next observation
++n_cases ;
} // For all cases
```

Example of Morlet Wavelet Series Generation

This section shows the DEEP.LOG file entries produced by a Morlet wavelet series generation. Prices from the S&P 100 Index OEX were transformed. The specifications shown next should be self-explanatory. But note that because a lag of 30 was used with a width of 40, the user is warned that the filter is compromised. Also, note that (Real_0, Imag_0) is the transformed value for the most recent source series observation, with the other variables being lagged values.

```
Reading series file D:\DEEP\TEST\OEX_TRAIN.TXT for Morlet
variables
  The log of the series is differenced to create the target;
  log of series for source
  Data is extracted from column 2
  Window length is 5 (gives twice this many predictors)
  Shift (spacing of adjacent cases in series) is 1
  Period is 20 samples
  Width each side of center is 40 samples
  Lag from current to center is 30 samples
    WARNING - Because the lag is less than the width,
              the filter will be compromised.
  Target tails are trimmed by 5.000 percent
  Targets are multiplied by 1.000
  A single target variable is used for a prediction model
  No header record is skipped
```

```
Morlet series target trimming is as follows...
Variable      Old min      New min      New max      Old max

 Lead_1      -0.23540     -0.01744      0.01720      0.10151

Basic statistics after any trimming...

Variable         Mean         StdDev          Min            Max
 Real_4      0.00006      0.00781      -0.06202      0.05355
 Imag_4     -0.00009      0.00781      -0.04440      0.06647
 Real_3      0.00005      0.00781      -0.06202      0.05355
 Imag_3     -0.00008      0.00781      -0.04440      0.06647
 Real_2      0.00005      0.00780      -0.06202      0.05355
 Imag_2     -0.00008      0.00781      -0.04440      0.06647
 Real_1      0.00005      0.00780      -0.06202      0.05355
 Imag_1     -0.00008      0.00781      -0.04440      0.06647
 Real_0      0.00004      0.00779      -0.06202      0.05355
 Imag_0     -0.00008      0.00781      -0.04440      0.06647
 Lead_1      0.00034      0.00876      -0.01744      0.01720
Lead_Pos     0.52794      0.49922       0.00000      1.00000
Lead_Neg     0.47206      0.49922       0.00000      1.00000
```

Path in an XY Plane

Many applications involve a time series of (X,Y) points that describe a path over a plane. Consider the following:

- A person writes something or draws a figure on a touchscreen using their finger or a stylus. Sampling the path at equal time intervals or equal distance intervals gives an (X,Y) series that can be used to identify what was drawn.
- You are in a military vehicle and observe another vehicle in front of you by means of a forward-looking infrared sensor. The path traced by the perimeter of that vehicle's IR signature defines a closed loop of (X,Y) points that can be used to identify the type of vehicle you are encountering.

- You trace the outline of a cell taken from a biopsy. The perimeter of that cell describes an (X,Y) path whose behavior may help decide if it is benign or malignant.
- You have a 2-D sensor attached to a machine's gearbox. The (X,Y) path defined by the vibration of the gearbox may indicate whether failure is immanent.

In any of these applications we may be able to use the raw (X,Y) data as inputs to a model. However, much practical experience has shown that computing the Fourier transform of the path almost always provides superior performance, especially if the complex values of the transform are used as inputs to a complex-domain neural network.

Normalization for Invariance

If you apply a Fourier transform to the raw path data, there are four sorts of normalization that you may want to apply after transforming in order to achieve invariance against several types of variation.

- *Location (centering)*: In the vast majority of applications you don't want the position of the complete path within the XY plane to impact results. For example, if a person writes a word on a touchscreen, whether he writes it near the top of the screen or near the bottom of the screen would, in most cases, be irrelevant to your decision-making process. Location invariance is trivial to achieve: just ignore the Re[0] and Im[0] components of the transform, which are the mean position of the path in the two dimensions. All other components are unaffected by constant offset in any direction.
- *Scale*: In most applications, the size of the path (change in scaling equally in both dimensions) would be irrelevant. Whether a person writes a word large or small on a touchscreen is probably of no concern. The distance of an infrared sensor from its target impacts the size of the image, which would usually be irrelevant. Scale invariance is fairly easy to obtain, especially in the case of perimeter tracing. Examine the magnitude of the first and last Fourier coefficients, those for [1] and [-1]. (Recall that the -*k* coefficient is

at index n–k after a fast Fourier transform.) These are the positive and negative fundamental frequencies. Find whichever has greater magnitude (it will be [1] if the object is traced clockwise and [-1] if traced counterclockwise) and divide all coefficients by that magnitude. An alternative application-dependent scaling method may have to be devised if the path is not a closed loop or if the path crosses itself (yikes!) at some point.

- *Starting point*: If you are tracing a closed-loop perimeter, in most cases you want your results to be independent of where you begin tracing so you don't have to devise a likely awkward method for consistently starting the trace from the same location. This begins with the same decision as scaling: determine whether the [1] or the [-1] coefficient is greater. Compute the phase of that coefficient and adjust all coefficients so that this phase is zero. This is equivalent to revising the starting point in such a way that the phase of the dominant fundamental frequency is zero. This is easily achieved by expressing the Fourier coefficients as (Magnitude, Phase) instead of (Real, Imaginary). A change in starting point does not affect any magnitudes. But an angular change of θ in the starting point impacts every Fourier coefficient by changing its phase by $f\theta$, where f is the frequency of the coefficient, with $-n/2 < f <= n/2$. Thus, we simply find the phase θ of the [1] or [-1] coefficient, whichever has greater magnitude, and subtract $f\theta$ from the phase of every coefficient. This zeros the phase of the dominant fundamental and adjusts all other phases accordingly. We thereby get the same Fourier transform regardless of where the perimeter trace begins.

- *Rotation*: Some applications require that we get the same Fourier transform regardless of the orientation of the perimeter in the XY plane, while for other applications the orientation is critical information. We'll discuss this in a moment. But for now, understand that instilling rotational invariance is extremely difficult, theoretically complicated, and fraught with risks. The DEEP program does not have this option, and in most cases you would be best off finding an application-specific way to achieve this, such as aligning the major axis in a predefined direction. But if you are interested in pursuing a Fourier transform–related method for producing rotational invariance, you can find an accessible discussion and source code in my book *Signal and Image Processing with Neural Networks*. It is out of print but still available used. An excellent overview of a variety of techniques is in the 1980 paper "An Efficient Three-Dimensional Aircraft Recognition Algorithm Using Normalized Fourier Descriptors" by Timothy Wallace and Paul Wintz, appearing in *Computer Graphics and Image Processing* Vol. 13.

Rotational invariance demands a bit more discussion. Suppose you have an object's perimeter traced on an XY plane. You sample points along that perimeter and apply a Fourier transform. Now suppose you rotate that object on the plane, trace its perimeter, and transform again. Is it important that you end up with the same transform? After all, it's the same object. Or is it important that you obtain a different transform, implying that not only is the shape of the object important but its orientation as well? That depends on the application.

Suppose your "objects" are letters written out with a thick marker, and you trace the outer perimeter of the letter. Then certainly orientation is important. You want M and W to have very different Fourier signatures. Or suppose you have traced the outline of a nearby vehicle using a FLIR sensor, and you want to identify the vehicle by the shape of its perimeter. You know that the bottom of the vehicle is on the ground, so the orientation is a key component of the identification process. In both of these cases you would not want rotational invariance.

Now suppose you are looking at a vehicle from above, perhaps from high- resolution satellite photography. In this case, the orientation of the vehicle is random and unrelated to the identity of the vehicle. If you got a significantly different Fourier signature for every possible orientation of the same vehicle, you would be in trouble. You want a unique signature for every vehicle, regardless of its orientation. For this application you would need rotational invariance.

Finally, suppose you are tracing the perimeter of a cell from a biopsy, and you know that benign versions of this cell tend to have a smooth perimeter, while malignant versions of this cell have a somewhat more convoluted perimeter. This property is not related to orientation at all. It's not as if malignant cells have a fixed shape that may occur in any orientation, like a vehicle. Rather, this is just a property of the perimeter that can occur in an infinite number of ways. So, in this application, it's a toss-up whether rotational invariance would help or hurt. You might want to try both and see whether either gives superior results, but chances are they would have similar performance.

The bottom-line question is this: does the orientation of the object in the XY plane provide valuable information (as in character identification), or is it a troubling distraction (as in vehicle identification from above)? If the answer is not clear, then chances are that rotation invariance does not matter much.

Pseudocode for XY Plane Series Processing

In this section we present a highly abbreviated algorithm in C-like pseudocode for implementing the generation of predictors and targets by means of a Fourier transform of a series of points in the XY plane. You can find this "code" in the file SERIES.TXT.

The user of the DEEP program chooses from one of four options for how variables will be computed for the database and which of these will go into the default predictor list:

1) *Raw XY*: The raw XY points go into the database and the default predictor list.
2) *RAW Fourier*: The complete set of Fourier transform coefficients go into the database and the default predictor list.
3) *Fourier, location normalized*: All Fourier transform coefficients go into the database. All except Re[0] and Im[0] go into the default predictor list.
4) *Fourier, scale/start normalized*: The Fourier transform is scale-normalized by dividing all coefficients except [0] by the magnitude of the dominant fundamental, [1] or [-1]. The phase of the dominant fundamental is set to zero and the phase of all other coefficients adjusted accordingly. All of these coefficients are written to the database. (The dominant fundamental will have its real part 1.0 and its imaginary part 0.0 for all cases, which lets the user determine which fundamental was dominant if automated choice is selected.) All coefficients except [0] and the dominant fundamental go into the default predictor list.

To be perfectly clear, as well as to provide an outline for writing write your own code, here is the pseudocode for assigning variable names to the database and the default predictor list. We keep two index counters. The counter n_vars indexes database variables, while n_pred indexes default predictor variables. This loop is divided into two large sections. The first section handles X or the real Fourier coefficient. The second section handles Y or the imaginary coefficient. All variables go into the database, which is the first thing done in each section.

```
n_pred = 0 ; // Will index default inputs
n_vars = 0 ;

for (i=0 ; i<n ; i++) { // For all XY pairs

  if (Fourier variables)
    Put name "Real_i" in dataset name array at slot n_vars ;
  else
    Put name "X_i" in dataset name array at slot n_vars ;
```

For raw data, all XY points go into the default predictor list. But for all Fourier variables, we put in coefficients only up through the maximum frequency as specified by the user. Recall that a fast Fourier transform leaves the coefficient for f=$-k$ at location n–k.

```
  if (Raw data) { // Raw XY points, no Fourier
    Put name "X_i" in default predictor list at slot n_pred
    ++n_pred ;
    }

  else if (i <= maxfreq || (n-i) <= maxfreq) {   // Use both pos and neg frequencies
```

If the user requested raw Fourier variables, all coefficients go into the default predictor list.

```
    if (Raw Fourier) {
      Put variable "Real_i" in default predictor list at slot n_pred
      ++n_pred ;
      }
```

If the user requested location-normalized Fourier variables, all coefficients except [0] go into the default predictor list.

```
else if (Location-normalized Fourier) {
  if (i > 0) { // If centered, we omit [0] terms
    Put variable "Real_i" in default predictor list at slot n_pred
    ++n_pred ;
    }
  }
```

If the user requested scale/start normalized Fourier, we omit [0] because nobody would want to do scale/start normalization without also doing location normalization. The dominant fundamental will also be omitted, and this is determined by whether the perimeter tracing proceeded clockwise (Mag[1] > Mag[-1]) or counterclockwise (Mag[-1] > Mag[1]). This, of course, must either be explicitly stated by the user or tested before names are assigned.

```
else if (Scale/start normalized Fourier) {
  if (i > 0) {
    if (is_clockwise) {
      if (i != 1) {
        Put variable "Real_i" in default predictor list at slot n_pred
        ++n_pred ;
        }
      }
    else {
      if (i != n-1) {
        Put variable "Real_i" in default predictor list at slot n_pred
        ++n_pred ;
        }
      }
    }
  }
}

++n_vars ;
```

The second half of the loop handles the Y or imaginary terms. It will be shown here in its entirety so that you can grasp the complete flow.

```
if (Fourier variables)
  Put name "Imag_i" in dataset name array at slot n_vars ;
else
  Put name "Y_i" in dataset name array at slot n_vars ;

if (Raw data) { // Raw XY points, no Fourier
  Put name "Y_i" in default predictor list at slot n_pred
  ++n_pred ;
  }

else if (i <= maxfreq || (n-i) <= maxfreq) { // Use both pos and neg frequencies
  if (Raw Fourier) {
    Put variable "Imag_i" in default predictor list at slot n_pred
    ++n_pred ;
    }
  else if (Location-normalized Fourier) {
    if (i > 0) { // If centered, we omit [0] terms
      Put variable "Imag_i" in default predictor list at slot n_pred
      ++n_pred ;
      }
    }
  else if (Scale/start normalized Fourier) {
    if (i > 0) {
      if (is_clockwise) {
        if (i != 1) {
          Put variable "Imag_i" in default predictor list at slot n_pred
          ++n_pred ;
          }
        }
      else {
        if (i != n-1) {
          Put variable "Imag_i" in default predictor list at slot n_pred
          ++n_pred ;
          }
```

```
          }
        }
      }
    }

  ++n_vars ;
  }
```

Now that name assignment is clear, we can present the pseudocode for computing the Fourier variables and outputting them to the database. In this code, *n* is the number of XY points. We allocate working memory and create an FFT object. The FFT code is in MRFFT.CPP.

```
n = Number of XY points ;

xr = [Allocate memory for n doubles] ;
xi = [Allocate memory for n doubles] ;
mag = [Allocate memory for n doubles] ;
phase = [Allocate memory for n doubles] ;
fft = new FFT ( n , 1 , 1 ) ;
```

The main loop starts here. The *n* original XY points are in record as we begin, as (X,Y) pairs. When we are done, record will contain the *n* Fourier variables as (Real, Imaginary) pairs. We copy the XY pairs and apply the fast Fourier transform. We have to divide by *n* because cpx() does not do this for a forward transform.

```
for all cases {
  Get all (X,Y) pairs for this case into record, which is 2n long

  for (i=0 ; i<n ; i++) {
    xr[i] = record[2*i] ;
    xi[i] = record[2*i+1] ;
    }

  fft->cpx ( xr , xi , 1 ) ; // Transform to frequency domain

  for (i=0 ; i<n ; i++) {   // fft sums, does not do this division
    xr[i] /= n ;
    xi[i] /= n ;
    }
```

If the user requested raw Fourier coefficients or location normalized coefficients (which simply means that we omit [0] from predictors), we just copy the coefficients to record. That's it.

```
if (Raw Fourier or centered) {      // If centered will omit [0] from default predictors
  for (i=0 ; i<n ; i++) {
    record[2*i] = xr[i] ;
    record[2*i+1] = xi[i] ;
    }
  }
```

If the user requested scale/start normalization, things are more complicated. We compute the magnitude and phase of every component.

```
else { // Normalize scale and phase
  for (i=0 ; i<n ; i++) {
    mag[i] = sqrt ( xr[i] * xr[i] + xi[i] * xi[i] ) ;
    if (mag[i] > 1.e-30)
      phase[i] = atan2 ( xi[i] , xr[i] ) ;
    else
      phase[i] = 0.0 ;
    }
```

The user has the option of directly specifying that the perimeter tracing was clockwise or counterclockwise. The user had better be right if this is the case, or the scaling and normalization will be seriously incorrect. The user can also choose to let the program determine the direction of tracing, which is the best choice in most situations.

```
if (Force clockwise)
  is_clockwise = 1 ;
else if (Force counter-clockwise)
  is_clockwise = 0 ;
else
  is_clockwise = mag[1] >= mag[n-1] ;
```

We first normalize the scale. The dominant fundamental determines the scale factor. As long as we are doing this, we might as well get the dominant fundamental's phase as well, as we will need it for starting-point normalization.

```
// Normalize the scale
if (is_clockwise) {
  scale = mag[1] ;
  theta = phase[1] ;
  }
else {
  scale = mag[n-1] ;
  theta = -phase[n-1] ;
  }

if (scale < 1.e-30) // Will never happen in valid app
  scale = 1.e-30 ;
for (i=1 ; i<n ; i++)
  mag[i] /= scale ;
```

We now normalize the phase, separately for the positive and negative frequencies. For the positive frequencies we go up through the Nyquist frequency, while for negative frequencies we go up to but not including the Nyquist frequency. The number of negative frequencies depends on whether n is even or odd.

```
// Normalize the phase for positive frequencies
for (i=1 ; i<=n/2 ; i++)
  phase[i] -= i * theta ;
// Normalize the phase for negative frequencies
k = (n % 2) ? (n/2) : (n/2-1) ;
for (i=1 ; i<=k ; i++)
  phase[n-i] += i * theta ;
```

Models generally do not like working with magnitude and phase, so we convert back to (Real, Imaginary) pairs and put them back into record.

```
// Recompute the normalized Fourier coefficients
for (i=0 ; i<n ; i++) {
  xr[i] = mag[i] * cos(phase[i]) ;
  xi[i] = mag[i] * sin(phase[i]) ;
  }
```

```
// Put the coefs back into the database
for (i=0 ; i<n ; i++) {
   record[2*i] = xr[i] ;
   record[2*i+1] = xi[i] ;
   }
} // Normalize scale and starting point

Advance to next case
} // For all cases
```

Example of XY Plane Series Processing

Some types of cancer cells can be distinguished from their benign counterparts by having somewhat more irregular borders. And there are mechanical equivalents as well; sometimes gears or pulleys can be identified as acceptable versus defective according to their eccentricity. With this inspiration I created an artificial dataset consisting of circles whose perimeter has random waviness with random period and of a user-assignable degree. The degree of departure from being perfectly round defines their membership in one of two classes.

Two datasets, each consisting of 1,000 cases, were created so that one could act as a training set and the other as a test set. The perimeter was sampled at 100 points. To prevent Fourier-based normalization from creating an unfair advantage over using the raw XY points as predictors, the following preprocessing was done prior to presenting the data to the DEEP program:

- The center of rotation of all examples is set at the same point, the origin (0,0). This prevents location normalization from providing an advantage to Fourier processing.
- The mean radius of all examples is the same. This prevents scale normalization from providing an advantage to Fourier processing.
- The starting point of each example is almost the same, randomly varying within a range of +/–9 degrees. This is far tighter than any practical raw-data standardization could accomplish and hence is a conservative approach. This prevents starting-point normalization from providing a significant advantage to Fourier processing.

To eliminate any effects because of unequal training, the classification model employed is just direct predictor-to-class linear regression, achieved by selecting an RBM/Supervised model with 0 unsupervised layers and 1 supervised layer, the output layer.

When the raw data points are used as predictors, the following results appear in the log file:

```
Reading XY file D:\TEST\BLOB_TRN.TXT with 100 points and 2
classes
User requests using raw (X,Y) points as predictors
Successfully read 1000 cases with 202 variables

Confusion matrix... Row is true class, column is predicted
class. In each set of three rows for a true class, the first
row is the count, the second row is the percent for that row
(true class) and the third row is the percent of the entire
dataset.

          1       2

  1     375     125
      75.00   25.00
      37.50   12.50

  2     117     383
      23.40   76.60
      11.70   38.30

Total misclassification = 24.2000 percent

Reading XY file D:\TEST\BLOB_TST.TXT with 100 points and 2
classes
User requests using raw (X,Y) points as predictors
Successfully read 1000 cases with 202 variables
```

```
Test results...

           1        2

  1      300      200
       60.00    40.00
       30.00    20.00

  2      212      288
       42.40    57.60
       21.20    28.80

Total misclassification = 41.2000 percent
```

Observe that the classification quality leaves a lot to be desired. The test set is almost random in its classification.

When the experiment is repeated using scale/start-normalized Fourier descriptors, the results shown next are obtained. There's no contest! When classifying by a perimeter trace, Fourier descriptors are hard to beat.

```
Reading XY file D:\TEST\BLOB_TRN.TXT with 100 points and 2
classes
User requests using scale/start-normalized Fourier transform
as predictors
User requests automatic rotation determination
Maximum frequency = 50
Successfully read 1000 cases with 202 variables

Confusion matrix... Row is true class, column is predicted
class. In each set of three rows for a true class, the first
row is the count, the second row is the percent for that row
(true class) and the third row is the percent of the entire
dataset.
```

```
          1        2

  1      500        0
      100.00     0.00
       50.00     0.00

  2        0      500
        0.00   100.00
        0.00    50.00

Total misclassification = 0.0000 percent

Reading XY file D:\TEST\BLOB_TST.TXT with 100 points and 2
classes
User requests using scale/start-normalized Fourier transform
as predictors
User requests automatic rotation determination
Maximum frequency = 50
Successfully read 1000 cases with 202 variables

Test results...
          1        2

  1      500        0
      100.00     0.00
       50.00     0.00

  2        0      500
        0.00   100.00
        0.00    50.00

Total misclassification = 0.0000 percent
```

CHAPTER 3

Image Preprocessing

An enormous variety of algorithms exist for preprocessing images for presentation to a model. This chapter will discuss only one such algorithm, though it is an important one whose computational details are often glossed over in other references. Here we will downplay the deep theory, which is widely available, and focus on the practical implementation details, which are not so widely available.

The Fourier Transform in Two Dimensions

We have already studied the Fourier transform in one dimension. (You may want to review the material starting on page 41.) It is often useful for time-series analysis. It does have the property that it fuses all of the information in the entire series into a set of numbers that encompasses the entire time period. Time-dependent changes in the frequency content of the signal are lost for all practical purposes. That led us to study the Morlet wavelet transform. But sometimes such information fusion is appropriate for an application. If the series is such that we can assume that its frequency content is effectively constant across its extent, then the ordinary Fourier transform is acceptable. The same holds true for image processing. In this section we will present the two-dimensional discrete Fourier transform (DFT) as applied to individual images.

The one-dimensional DFT is well known and will not be repeated here. Its somewhat more complicated generalization to two dimensions is shown in Equation 3-1, the forward transform, and Equation 3-2, the inverse transform.

$$H\left(f_x, f_y\right)=\sum_{x=0}^{n_x-1}\sum_{y=0}^{n_y-1} h(x,y)e^{\left[i\frac{2\pi x f_x}{n_x}\right]}e^{\left[i\frac{2\pi y f_y}{n_y}\right]} \tag{3-1}$$

$$h(x,y)=\frac{1}{n_x n_y}\sum_{f_x=0}^{n_x-1}\sum_{f_y=0}^{n_y-1} H\left(f_x, f_y\right)e^{\left[-i\frac{2\pi x f_x}{n_x}\right]}e^{\left[-i\frac{2\pi y f_y}{n_y}\right]} \tag{3-2}$$

T. Masters, *Deep Belief Nets in C++ and CUDA C: Volume 2*, https://doi.org/10.1007/978-1-4842-3646-8_3

Similar to the case in one dimension, f_x in these equations takes on n_x different integer values. Each value is a frequency. It is the number of complete cycles per the horizontal extent of the image. The maximum horizontal frequency that can be resolved without aliasing, the Nyquist frequency, is $n_x / 2$. This corresponds to a period of two pixels. So that array indexing is easy, we generally use the simpler but less efficient convention of letting f_x range from zero through $n_x - 1$ when programming the transform. Values of f_x beyond the Nyquist limit correspond to negative frequencies. For the remainder of this section, we will use whichever of these two alternative interpretations (negative, or positive past Nyquist) is most convenient at the time.

The same interpretation applies to the vertical frequencies. The maximum vertical frequency is $n_y / 2$ cycles per the vertical extent, and f_y will take on n_y different integer values.

The two-dimensional DFT of Equations 3-1 and 3-2 looks long and complicated, but really it is not. Each of the two exponential terms is a *cosine + i sine* wave. One term varies along the *x* direction at a frequency determined by f_x. The other term varies in the *y* direction at a frequency determined by f_y. This is just two separate transforms, one operating in the horizontal direction and the other operating in the vertical direction. Image variation in other directions will have vertical and horizontal components that will be picked up according to their relative strengths.

Let's pursue this line of thought a little more so that we can better understand the nature of the transform. Suppose that $f_x = 0$. The first exponential term in the transform equation will then equal one for all values of *x*, so it can be ignored. It can be seen that $H(0,f_y)$ is the sum over all columns (*x* values) of ordinary Fourier transforms. Each transform is the one-dimensional transform of the rows of a column. In other words, $H(0,f_y)$ represents the overall variation in the image due to up-down ripples at a frequency of f_y cycles per vertical extent. It can similarly be seen that $H(f_x, 0)$ accounts for the variation in a horizontal direction.

The previous property can be generalized. To keep things simple, assume that the image is square so that $n_x = n_y = n$. Recall that $e^a e^b = e^{a+b}$. Express the domain variables of the transform in polar coordinates, as shown in Equation 3-3.

$$\left(f_x, f_y\right) = \left(k \cos\theta, k \sin\theta\right) \tag{3-3}$$

Now consider the product of the exponential terms in the transform. Apply the multiplication rule and express the frequencies in polar coordinates. The product term can then be written as shown in Equation 3-4.

$$e^{\left[i\frac{2\pi x f_x}{n}\right]} e^{\left[i\frac{2\pi y f_y}{n}\right]} = e^{\left[\frac{2\pi i}{n}(x\ k\ \cos\ \theta + y\ k\ \sin\ \theta)\right]} \tag{3-4}$$

This expression tells us something very useful. It tells us that image variation that is at a frequency of k cycles per unit distance and in a direction of θ will be detected by H(k cos θ, k sin θ). Mathematically inclined readers who would like a more rigorous derivation are given the following exercise. Write the expression for a wave at a frequency of k cycles per unit distance and in a direction of θ. Use trigonometry to compute its number of cycles per unit distance in the horizontal and vertical directions and then show the effect of each of these variations on the quantity expressed in Equation 3-1. The result will be the same. I like the more intuitive approach of considering each transform term as representing a projection on an individual wave traveling in a particular direction at a particular frequency.

In actuality, the situation is slightly more complicated than the simple formula just described. The representation is orthogonal only on a discrete lattice in Cartesian coordinates. Therefore, parameterizing the transform in terms of polar coordinates can be confusing. Also, issues of side lobes have been sidestepped. They will be discussed in more detail on page 90. But the approximation is very good, and the intuitive appeal is excellent. In practice, this interpretation is totally serviceable.

There is one subtlety to note in regard to this formula. The angle of image variation, θ, is assumed to be measured clockwise from a right-hand direction. This is because rows increase going downward, just the opposite of Cartesian coordinates. Most computer programs, including the one presented in the next section, transform in this top-to-bottom direction. This should be kept in mind if explicit directions are important.

An example may clarify the interpretation of the transform coefficients. Suppose we know enough about the physics of our application to know that we are particularly interested in detecting variation that occurs downward and to the right, at an angle of 30 degrees clockwise from directly right. Also suppose that this variation will typically have a frequency of 6 cycles per the length of a side of the transformed square. We must multiply 6 times the cosine and sine of 30 degrees to get the values of f_x and f_y that will be most sensitive to this variation. In particular, $f_x = 5.2$ and $f_y = 3$. Since the DFT is most often computed using integer values for the frequencies (although it does not have to be), we find that H(5, 3) will be the transform coefficient most sensitive to the variation of interest. Side lobes will cause some of the energy at that frequency to appear in other terms, but most of it will go into that one. Also note that later in this chapter the restriction to square areas will be dropped.

Integer values for the frequencies are not mandatory. It is just that “fast” algorithms are difficult or impossible for arbitrary real values. In the unusual case that an application involves a few highly specific frequencies, Equation 3-1 can be explicitly evaluated for the frequencies of interest using brute-force computation. It will be slow. But if only a few coefficients need to be found, the speed may be commensurate with that obtained by using the fast DFT to compute all of the coefficients and then discarding most of them! And the increased accuracy obtained by zeroing in on the exact frequencies may be worth the effort.

When the image is real, there is an important symmetry in its two-dimensional Fourier transform. The transform value at any (f_x, f_y) point is the complex conjugate of the transform value at the point reflected about both axes. This is made explicit in Equation 3-5.

$$H\left(f_x, f_y\right) = \bar{H}\left(-f_x, -f_y\right) \tag{3-5}$$

You might have been alerted to this symmetry in the example given on the prior page. It was stated that the variation was to the lower right at an angle of 30 degrees. But that is essentially the same thing as variation to the upper left, at an angle of 30 + 180 = 210 degrees. Using the same formula leads us to H(–5, –3). The complex conjugate comes about from the sine wave being an odd function and hence undergoing a sign reversal when the direction reverses. Mathematicians would derive Equation 3-5 by flipping the signs of f_x and f_y in Equation 3-1 and seeing what happens. (The cosine component is even, so it is unchanged. The odd sine term flips its sign.)

The implication of this symmetry is that only about $n_x * n_y / 2$ complex terms are needed to completely describe the transform. This quantity should not be surprising. The original block of data consisted of $n_x * n_y$ real numbers. If the transform has half as many complex numbers (each equivalent to a pair of real numbers), then information is conserved, and all is well with the world.

To minimize the amount of data presented to the neural network, it is important to understand the symmetry in full detail. My habit is to compute the transform for all n_x values of f_x, corresponding to both positive and negative horizontal frequencies but only compute the $n_y / 2 + 1$ values of f_y that correspond to nonnegative vertical frequencies. That gives a few too many numbers. We need only about $n_x * n_y / 2$, but we are computing $(n_y / 2 + 1) * n_x$ numbers. Where did the extras come from? Look back at Equation 3-5. When $f_y = 0$, that equation tells us that the n_x terms corresponding to the positive and negative horizontal frequencies contain redundancy. The terms corresponding to negative frequencies are the complex conjugates of the positive-frequency terms. The same is true at the vertical Nyquist frequency, $f_y = n_y / 2$, since the Nyquist frequency has no sign. Thus, when $f_y = 0$ or $f_y = n_y / 2$, we can ignore the terms corresponding to negative values of f_x. Those are the extras.

Let's pursue this symmetry issue just a little longer. Many readers have undoubtedly taken offense at the quantity of independent transform terms being said to be *about* $n_x * n_y / 2$. We need to know *exactly* how many there are because that determines the number of input neurons in the network. Also, can we reconcile the transform with an input containing $n_x * n_y$ real numbers so that conservation of information is strict? We do so now.

For each value of f_y from 1 through $n_y / 2 - 1$, all n_x values of f_x will produce fully complex numbers. This gives us $(n_y / 2 - 1) * n_x$ complex numbers (network inputs) so far. For $f_y = 0$, we have $n_x / 2 + 1$ numbers that are not redundant, but the first (zero) and last (Nyquist) are strictly real. The others are general complex. The same is true when $f_y = n_y / 2$, the vertical Nyquist frequency. Thus, the vertical zero and Nyquist sets contribute a total of $n_x + 2$ complex network inputs. Adding things up gives us a total of $n_x * n_y / 2 + 2$ complex inputs to the neural network. But four of those inputs, the zero and Nyquist terms in the horizontal and vertical directions, are strictly real. *Voila.* All is right in the world.

There is a standard pattern for storing the transform. The code presented in the next section uses this pattern, as do many other common programs. It is illustrated in Figure 3-1. Each transform term in that figure is represented by f_x, f_y (which is column, row). Studying this figure may clarify some of the preceding and subsequent discussions.

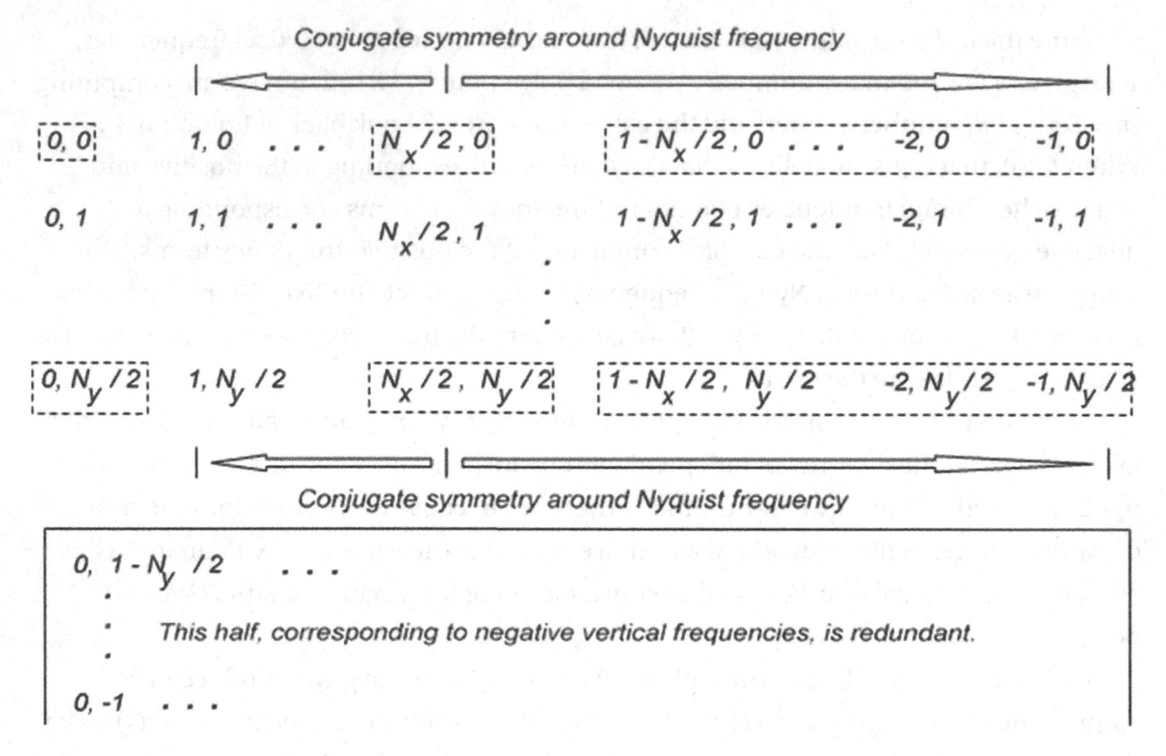

Figure 3-1. *Storage of 2-D FFT*

To avoid storing redundant information, only terms corresponding to nonnegative vertical frequencies are kept. The terms having negative vertical frequency, noted at the extreme bottom of the figure, are the complex conjugates of those corresponding to positive frequencies, reflected about both axes. The two groups of terms outlined by fine dotted lines, at the top right and bottom right, are redundant in that they are the complex conjugates of the terms to their left, reflected about the horizontal Nyquist term. However, they are still stored to keep things simple. Eliminating their storage would save very little space but would vastly complicate the situation. Finally, the four individual terms enclosed in boxes are strictly real. We know in advance that their imaginary components will always be zero.

One final point should be made. This entire discussion has focused on (and will continue to focus on) discarding negative vertical frequencies to avoid redundancy. Actually, there are three other possibilities. One could discard positive vertical frequencies, discard negative horizontal frequencies, or discard positive horizontal frequencies. In other words, the entire transform is a rectangle that is half redundant. We have chosen to keep the top half of the rectangle, shown in Figure 3-1. One could just as well keep the bottom half, left half, or right half. The choice is often a matter of personal preference. The method described here seems to be the most customary, but they are all correct.

Data Windows in Two Dimensions

Side lobes are every bit as important a consideration in two dimensions as they are in one dimension. Equation 3-1 and the surrounding discussion should be reviewed. This section presents a straightforward extension of those results to two dimensions.

It is tempting to think that if the Fourier transform is applied to the original raw image, then we have kept some sort of "purity" by not tampering with the data. In fact, quite the opposite is true. The Fourier transform assumes that the image extends to infinity in all directions. The net effect of this implicit assumption is that frequency components that do not lie precisely on the integer f_x and f_y values leak into significantly distant transform components. The leakage pattern is graphed as a function of f_x and f_y in Figure 3-2. This function is the amplitude of leakage relative to the amplitude of the center lobe. It has been truncated at 10 percent. Only one quadrant is shown since the others are symmetric. The worst part about this leakage function is that the height of the side lobes tapers off excruciatingly slowly. That means that serious leakage can occur across very large distances. This situation is virtually always intolerable.

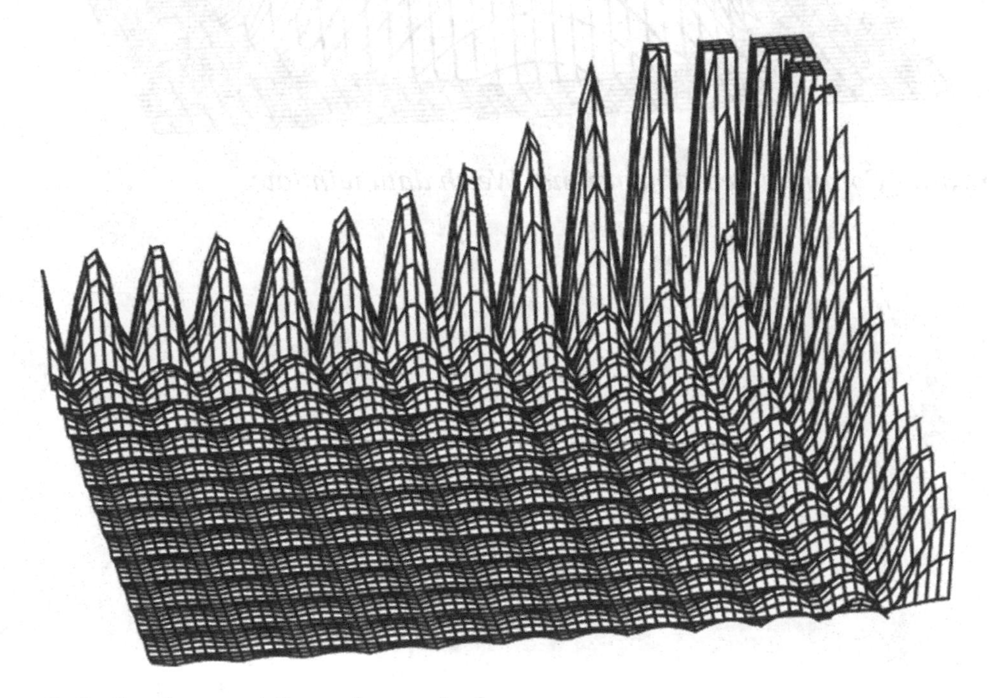

***Figure 3-2.** Leakage with no data window*

An excellent solution is to generalize the standard one-dimensional Welch data window, shown in Equation 3-1. The usual two- dimensional generalization is shown in Equation 3-6. This window is graphed in Figure 3-3, and its leakage is shown in Figure 3-4, with the height again truncated at 10 percent. Note the tremendous improvement over Figure 3-2.

$$w_{xy} = 1 - \left(\frac{x - 0.5(n_x - 1)}{0.5(n_x + 1)} \right)^2 - \left(\frac{y - 0.5(n_y - 1)}{0.5(n_y + 1)} \right)^2 \tag{3-6}$$

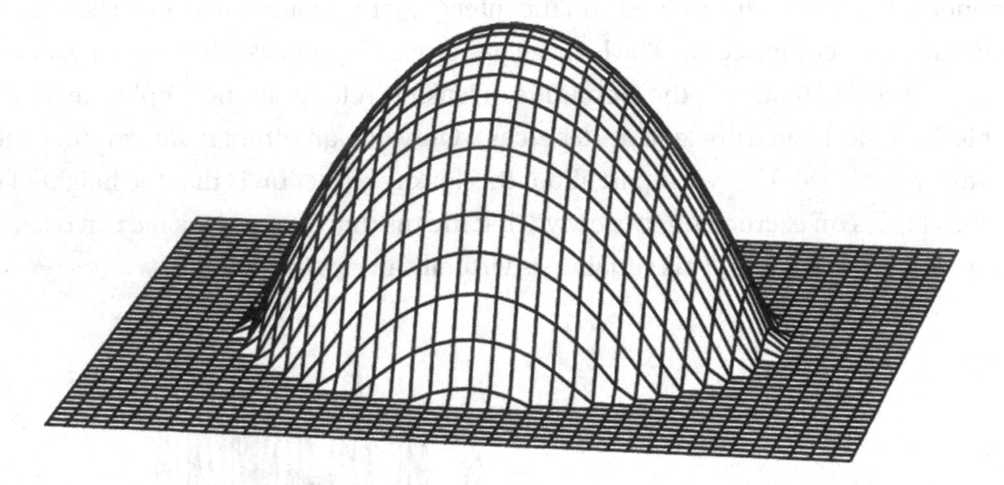

Figure 3-3. *Common two-dimensional Welch data window*

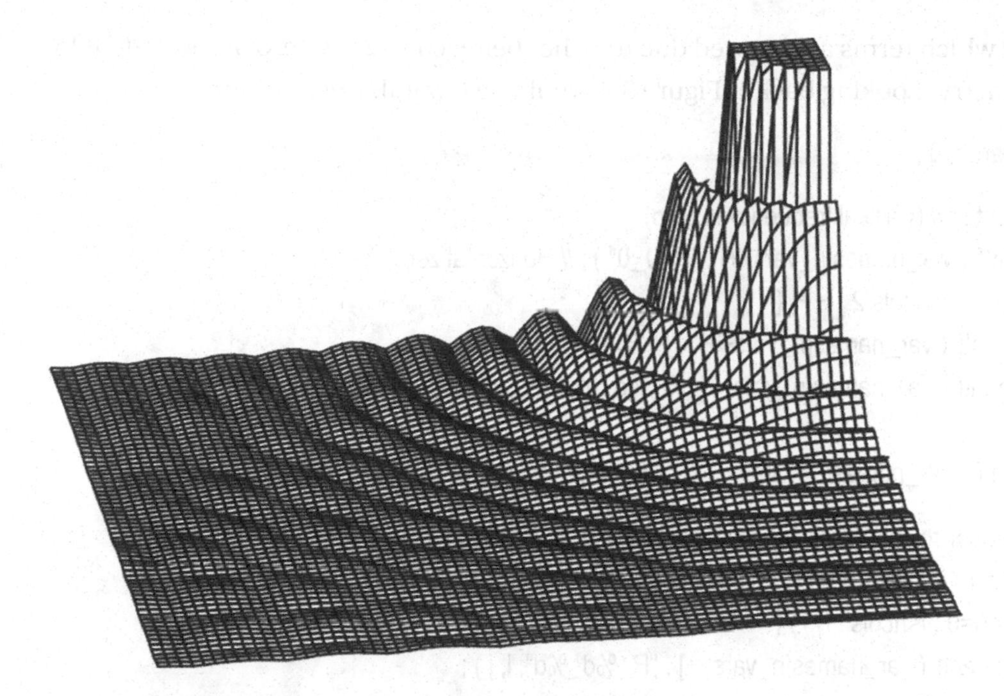

***Figure 3-4.** Leakage of the previous data window*

Several aspects of Equation 3-6 should be noted. The rectangle is n_y rows by n_x columns. The x and y indices run from 0 through $n_x - 1$ and $n_y - 1$, respectively. Most importantly, observe that the window as defined in that equation can become negative. We do not let that happen. If a computed value of w_{xy} is negative, set it equal to zero.

Code for the Fourier Transform of an Image

This section provides code snippets that illustrate computation of a two- dimensional Fourier transform of an image. We also show how to select the unique (nonredundant) values of the transform for presentation to a neural network or other model.

You'll see in Chapter 5 that the DEEP program names the transform variables using either the letter *R* for a real part or *I* for an imaginary part, followed by the horizontal frequency (never negative) and finally followed by the vertical frequency (also never negative). It is instructive to study the code for generating these names because it makes clear which terms are ignored due to either being constantly zero or redundant by symmetry. Looking back at Figure 3-1 would be helpful. Here is this code:

```
n_vars = 0 ;

// First row (vertical frequency is zero)
sprintf ( var_names[n_vars++] , "R_0_0" ) ; // Horizontal zero
for (i=1 ; i<ncols/2 ; i++) {
  sprintf ( var_names[n_vars++] , "R_%d_0", i ) ;
  sprintf ( var_names[n_vars++] , "I_%d_0", i ) ;
  }
sprintf ( var_names[n_vars++] , "R_%d_0", ncols/2 ) ; // Horizontal Nyquist

// Interior rows; All are full complex with no redundancy
for (j=1 ; j<nrows/2 ; j++) {
  for (i=0 ; i<ncols ; i++) {
    sprintf ( var_names[n_vars++] , "R_%d_%d", i, j ) ;
    sprintf ( var_names[n_vars++] , "I_%d_%d", i, j ) ;
    }
  }

// Last row (vertical frequency is Nyquist)
sprintf ( var_names[n_vars++] , "R_0_%d", nrows/2 ) ; // Horizontal zero
for (i=1 ; i<ncols/2 ; i++) {
  sprintf ( var_names[n_vars++] , "R_%d_%d", i, nrows/2 ) ;
  sprintf ( var_names[n_vars++] , "I_%d_%d", i, nrows/2 ) ;
  }
sprintf ( var_names[n_vars++] , "R_%d_%d", ncols/2, nrows/2 ) ; // Horizontal Nyquist
```

We are now ready to compute the Fourier transform. We will do all columns first and then all rows. In the most general case, the image need not be square, so we will need separate FFT objects. If we were transforming just one image, we could allocate the one for the columns, then delete it, and then allocate for the rows. But since we will usually be transforming numerous images, it makes more sense to allocate one of each for repeated use. The code for this class is in the module MRFFT.CPP available for free download from my web site.

```
fft_row = new FFT ( nrows , 1 , 1 ) ;
fft_col = new FFT ( ncols , 1 , 1 ) ;
```

- xr and xi are work arrays nrows * ncols long and will contain the computed transform when finished.
- rwork and iwork are work arrays nrows long.
- The nrows by ncols image is in pixels.

First we precompute the constants that will be needed for the Welch data window, as shown in Equation 3-6.

```
rcent = 0.5 * (nrows - 1) ; // Center for Welch data window
ccent = 0.5 * (ncols - 1) ;
rden = 0.5 * (nrows + 1) ; // And denominator
cden = 0.5 * (ncols + 1) ;
```

We transform the columns, one at a time, using xr and xi as work vectors to hold the columns. There is one aspect of the application of the Welch data window that is application-dependent and crucial. The window tapers to zero at the edges of the image, so we should make sure this value makes sense in the application. The MNIST images have a zero (or near zero) background with the digits having high tone values. (DEEP inverts them for non-Fourier training to make for a more conventional display). So, tapering to zero at the edges makes perfect sense. If instead the images were naturally high at the edges and low in the central information area, tapering the edges to zero would be terribly distorting, so the images must be inverted.

Here is the code for performing the transform:

```
for (icol=0 ; icol<ncols ; icol++) {            // Do each column
  cdist = (icol - ccent) / cden ;               // For Welch data window
  cdist *= cdist ;

  for (irow=0 ; irow<nrows ; irow++) {          // Copy this column to work vector
    rdist = (irow - rcent) / rden ;             // Simultaneously applying data window
    rdist *= rdist ;
    weight = 1.0 - cdist - rdist ;              // Equation 3-6
    if (weight < 0.0)
      weight = 0.0 ;

    rwork[irow] = weight * pixels[irow*ncols+icol] ; // Get the column
    iwork[irow] = 0.0 ;                         // Easy, slightly inefficient method
    } // For irow
```

```
    fft_row->cpx ( rwork , iwork , 1 ) ; // Transform rows of this column

    // We copy only zero through Nyquist; conjugate symmetric above vertical Nyquist
    // This is the upper (non-redundant) half of Figure 3-1
    for (irow=0 ; irow<=nrows/2 ; irow++) {
      xr[irow*ncols+icol] = rwork[irow] ;
      xi[irow*ncols+icol] = iwork[irow] ;
      }

    } // For icol

  // Now transform the rows
  for (irow=0 ; irow<=nrows/2 ; irow++)
    fft_col->cpx ( xr+irow*ncols , xi+irow*ncols , 1 ) ;
```

Astute readers may grumble about one thing done in this code. When transforming the columns, a full complex transform is done by putting zeroes in the imaginary parts of each column. There is a faster method that involves packing the entire column into a complex vector of half the length and then unpacking. The FFT class includes the member functions rv() and irv() for doing just this, and the method is documented in the source code. In my opinion, the added complexity is not worthwhile here, but you can easily make the modification.

After the code just shown has run, xr and xi contain the transform as laid out in the upper half of Figure 3-1. To reinforce the verbal explanations of what's redundant and what's not, we'll explore the code snippet that extracts the nonredundant data from this matrix. The order of extraction is the same as the order of variable naming discussed on page 94.

```
  k = 0 ;                              // Will index variables for this image

  // First row (vertical frequency is zero)
  dbptr[k++] = xr[0] ;                 // Horizontal frequency is zero
  assert ( fabs(xi[0]) < 1.e-6 ) ;     // By definition, imaginary part is zero

  for (i=1 ; i<ncols/2 ; i++) {        // The half past this is complex conjugate of this
    dbptr[k++] = xr[i] ;
    dbptr[k++] = xi[i] ;
    }
```

```
dbptr[k++] = xr[ncols/2] ;                   // Horizontal Nyquist
assert ( fabs(xi[ncols/2]) < 1.e-6 ) ;       // By definition, imaginary part is zero

// Internal rows, which are all full complex with no redundancy
for (j=1 ; j<nrows/2 ; j++) {
  for (i=0 ; i<ncols ; i++) {
    dbptr[k++] = xr[j*ncols+i] ;
    dbptr[k++] = xi[j*ncols+i] ;
    }
  }

// Last row (which is index nrows/2, the vertical Nyquist)
dbptr[k++] = xr[nrows/2*ncols] ;             // Horizontal frequency is zero
assert ( fabs(xi[nrows/2*ncols]) < 1.e-6 ) ;// By definition, imaginary part is zero

for (i=1 ; i<ncols/2 ; i++) {
  dbptr[k++] = xr[nrows/2*ncols+i] ;
  dbptr[k++] = xi[nrows/2*ncols+i] ;
  }

dbptr[k++] = xr[nrows/2*ncols+ncols/2] ;              // Horizontal Nyquist
assert ( fabs(xi[nrows/2*ncols+ncols/2]) < 1.e-6 ) ; // By def, imaginary part is zero

assert ( k == nrows * ncols ) ;               // This had better be true!
```

Displaying Generative Samples of Fourier Transforms

There are two issues to be considered when one displays generative samples of the Fourier transform of an image. First, each transform point is representative of two quantities, the real and imaginary components of the complex number associated with that point. Second, symmetry means that the number of complex points is about half of the number of points in the original image.

The first problem is the most critical. It is extremely rare for a display of just the real parts or just the imaginary parts to make visual sense. For this reason, most developers display either the power of each point, which is the sum of the squares of the real and imaginary parts, or the amplitude, which is the square root of the power. In a few applications, the log of the power is an appropriate quantity to display. The DEEP program displays the power.

But what about the fact that the power pools the real and imaginary parts, obscuring their relative contributions? In the vast majority of applications, this has little or no visual utility. But when it does, this relationship can be presented by color. Do not be tempted to do something simple like just using red for the real part and green for the imaginary part. This invariable produces a confusing display. The correct method is to use magnitude for brightness and use the arctangent function to compute the phase angle and then map the phase to a color scheme, which is also circular. This prevents ugly and cryptic discontinuities.

The symmetry issue is easier to deal with. The native data structure, shown in Figure 3-1, is generally agreed to be a poor way of displaying the transform. Instead, we usually place the zero frequency points in the center of the display, with increasing frequencies closer to the edge, and the Nyquist frequencies at the top row and left column. This way, the angular placement of unusually high power points will precisely correspond to the corresponding direction of overall variation in the image. As a useful exercise, review Equation 3-4 to see how this takes place.

The code to map the unique members of the native data structure of Figure 3-1 to a zero-centered display is surprisingly fussy. We actually map the order of variable naming discussed on page 94. The essential snippets are shown next. We will use ir and ic to refer to the row and column in the native structure, with the database in dptr, and use irow and icol to refer to the corresponding displayable image in synth_ptr.

The first section of the code sets the current row in the displayable image to be the center because we will begin with the vertical zero frequency, which goes in the center. The main loop runs the native data row index from zero through the vertical Nyquist, which covers the nonredundant components of the transform.

The code shown handles just one row, the zero vertical frequency. For this row, recall that the horizontal zero and Nyquist frequencies are pure real, and there is complex conjugate symmetry around the horizontal Nyquist. We use the index offset k to set two at a time according to this symmetry.

```
irow = image_rows / 2 ; // Zero vertical frequency goes in center of image

for (ir=0 ; ir<=image_rows/2 ; ir++) {   // Do entire upper half of Figure 3-1

  if (ir == 0) {                          // Zero vertical frequency
    icol = image_cols / 2 ;   // Will index columns of this row in synthetic image
    synth_ptr[irow*image_cols+icol] = *dptr * *dptr ; // Zero frequency is pure real
    ++dptr ; // Pure real, so increment once
```

```
// Interior of row, prior to Nyquist, is complex and symmetric
for (k=1 ; k<image_cols/2 ; k++) {
  synth_ptr[irow*image_cols+icol+k] = dptr[0] * dptr[0] + dptr[1] * dptr[1] ;
  synth_ptr[irow*image_cols+icol-k] = synth_ptr[irow*image_cols+icol+k] ;
  dptr += 2 ; // Full complex, so increment by 2 (real and imaginary)
  }

// Nyquist is pure real; goes in first column of image
synth_ptr[irow*image_cols+0] = *dptr * *dptr ;
++dptr ; // Pure real, so increment once
++irow ;
} // If ir == 0
```

The interior of the transform is a bit easier because all components are full complex with no symmetry across rows. However, as shown in Figure 3-1, we do have vertical symmetry around the Nyquist frequency. Copying this to the lower half of the visible display is not necessary because it is entirely redundant, but most developers agree that this copying makes for a more visually appealing display, both because it makes the display the same size and shape as the original image and because the symmetry can visually highlight important features.

When we do this copying, beware of two things. First, the zero horizontal frequency component stands alone, without symmetry. Second, the symmetry reverses the rows. (It also involves complex conjugation, which must be handled if phase is computed.)

The mapping begins with the zero horizontal frequency in the middle of the displayed image. As the native structure advances from zero, the displayed column also advances, but it must then wrap around to the first column when it reaches the end.

```
else if (ir < image_rows/2) { // Internal rows, between vertical zero and Nyquist
  icol = image_cols / 2 ;       // Will index columns of this row in synthetic image

  for (k=0 ; k<image_cols ; k++) {   // Entire row is complex
    synth_ptr[irow*image_cols+icol] = dptr[0] * dptr[0] + dptr[1] * dptr[1] ;
    // put redundant data in lower half for nice display

    if (icol == 0) // No symmetry for zero frequency; symmetry is around Nyquist
      synth_ptr[(image_rows-irow)*image_cols] = synth_ptr[irow*image_cols] ;
```

```
      else
        synth_ptr[(image_rows-irow)*image_cols+(image_cols-icol)] =
                                          synth_ptr[irow*image_cols+icol] ;
      if (++icol == image_cols)
        icol = 0 ;
      dptr += 2 ;
      }
    ++irow ;
    } // If ir < image_rows/2
```

We are almost done. The last step is to map the vertical Nyquist row to the top row of the displayed image. This is very similar to how we handled the first (zero vertical frequency) row.

```
  else {      // Vertical Nyquist goes in top row of image, reversed
    assert ( irow == image_rows ) ; // It won't really go here. Just checking.
    icol = image_cols / 2 ; // Will index columns of this row in synthetic image
    synth_ptr[0*image_cols+icol] = *dptr * *dptr ; // Zero frequency is pure real
    ++dptr ;

    // Interior of row, prior to Nyquist, is complex and symmetric around Nyquist
     for (k=1 ; k<image_cols/2 ; k++) {
       synth_ptr[0*image_cols+icol+k] = dptr[0] * dptr[0] + dptr[1] * dptr[1] ;
       synth_ptr[0*image_cols+icol-k] = synth_ptr[0*image_cols+icol+k] ;
       dptr += 2 ;
       }

    // Horizontal Nyquist is pure real; goes in first column of image
     synth_ptr[0*image_cols+0] = *dptr * *dptr ;
     ++dptr ;
     } // Vertical Nyquist

  } // For ir (rows of transform)
```

To illustrate the display of Fourier transforms, Figure 3-5 displays the power as computed with the preceding code for a cleanly written numeral 0 in the MNIST dataset. Notice the near symmetry. Similarly, Figure 3-6 is a cleanly written numeral 1, which has a slight clockwise slant from vertical. The dominant variation is, of course, at right angles to the line forming the numeral.

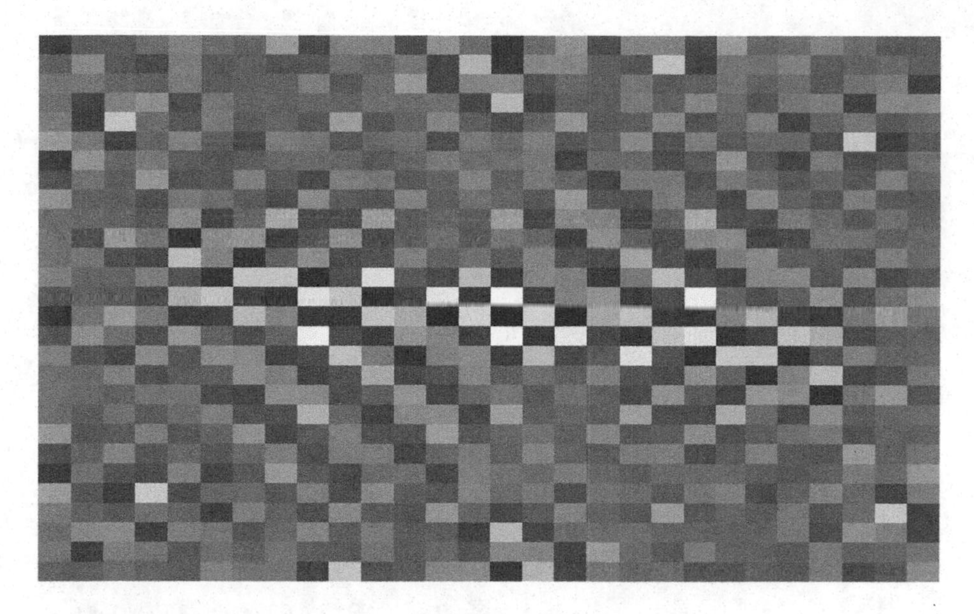

Figure 3-5. *Fourier transform of numeral 0*

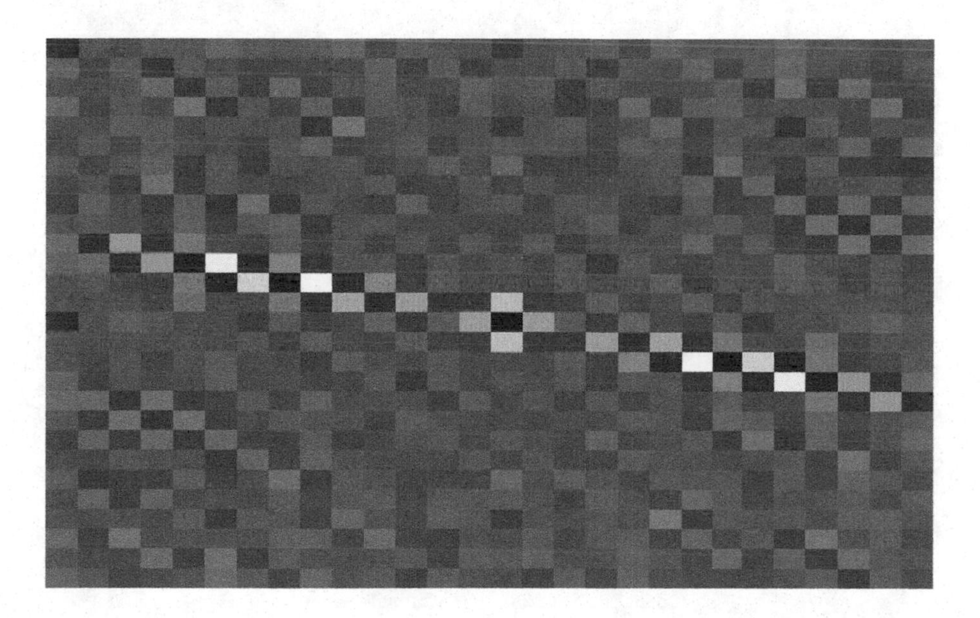

Figure 3-6. *Fourier transform of numeral 1*

CHAPTER 4

Autoencoding

Autoencoding as a means of creating deep belief nets predates the use of restricted Boltzmann machines (RBMs). This is perhaps because of their conceptual simplicity. The most basic autoencoder is an ordinary feedforward network that has a single hidden layer and is trained to reproduce its inputs. It's a prediction model in which the targets are the inputs. The idea is that if the hidden layer is in some sense relatively weak (perhaps by virtue of having few neurons or having limited weight magnitudes or some other form of regularization), this hidden layer will learn to encapsulate the "important" features of the training data, those that are most consistent and have highest information content. These significant features, which are defined by the activation pattern of the hidden layer, can then be used for classification or prediction, or they can be used as inputs to yet another autoencoder for further pattern extraction. An unlimited number of simple one-hidden-layer autoencoders can be stacked into a deep network.

Here are a few comments, comparisons, and contrasts between these two common methods for building deep belief nets, RBMs, and autoencoders:

- Both methods rely on discovering a hierarchy of features, ranging from primitive to complex. In image processing, the first layer may identify edges, the second layer may pick up small objects built from these edges, the third layer may discover more complex scene components composed of these small objects, and so forth.
- Both methods are easily amenable to greedy training, which means that deep networks are attainable without serious computational time constraints or numerical difficulties.
- Usually, autoencoding layers are faster (often tremendously faster) to train than RBM layers.

T. Masters, *Deep Belief Nets in C++ and CUDA C: Volume 2*, https://doi.org/10.1007/978-1-4842-3646-8_4

- Although RBMs have a strong tendency to reproduce inputs, this is not their primary goal, and reproduction is often far from perfect. Instead, the primary goal of an RBM is to encapsulate the statistical distribution of the training data. This difference can mean that one or the other is superior, depending on characteristics of the application.
- Because an RBM encapsulates the distribution of the training data, an RBM is said to be a *generative* model. This means that it can be persuaded to generate random samples from the distribution it has perceived as that from which the training data was drawn. Such samples of what the model has learned can be extremely informative in some applications, though useless in others. An autoencoder does not have this capability.
- Because an RBM incorporates a significant element of randomness in its training, it is inherently predisposed against overfitting. Even if the RBM contains more hidden neurons than it has inputs, a situation which in traditional modeling would be considered preposterous, it is unlikely to overfit. Naive autoencoders have no such advantage, so regularization must be imposed if overfitting is to be avoided. My own favorite method for autoencoders, which is *not* universally endorsed by experts, is to limit the number of hidden neurons in each layer and then train it until the cows come home. This encourages repeatability, which is always nice in an experimental situation. Other popular methods include stopping training early, limiting the magnitude of weights, stochastic gradient descent, and sparsity constraints. The first two alternatives can be implemented in DEEP by setting a small limit on training iterations or setting a large weight penalty.
- The choice of whether to build a deep belief net from RBMs or autoencoders is strongly application-dependent.
- Many of the "traditional" autoencoders described in the literature stay close to RBM approaches in that they deal with discrete-valued variables and/or probability distributions. The autoencoders described in this book take a very different approach: they ignore distributions, deal with unbounded continuous data, and can

operate in the complex domain. Thus, readers familiar with the usual presentations of autoencoders seen in the mainstream literature will find the approach taken here to be quite different, more limited in some ways and vastly more general in other ways.

Basic Mathematics of Feedforward Networks

A multiple-layer feedforward network is generally illustrated as a stack of layers of "neurons" similar to what is shown in Figure 4-1. This illustrates the most basic version, a network with a single hidden layer. For clarity, only a few of the connections appear; actually, every input neuron feeds every hidden neuron, and every hidden neuron feeds every output neuron. The bottom layer is the input to the network, what would be referred to as the independent variables or predictors in traditional modeling literature. The layer above the input layer is the first (and only, here) hidden layer. Each neuron in this layer attains an activation that is computed by taking a weighted sum of the inputs, plus a constant bias, and then applying a nonlinear function. Each hidden neuron in this layer will have a different set of input weights.

If there is a second hidden layer, the activations of each of its neurons is computed by taking a weighted sum of the activations of the first hidden layer plus a bias and applying a nonlinear function. This process is repeated for as many hidden layers as exist.

The topmost layer is the output of the network. There are many ways of computing the activations of the output layer, and several of them will be discussed later. For now let's assume that the activation of each output neuron is just a weighted sum of the activations of the neurons in the prior layer, plus a constant bias, without use of a nonlinear function.

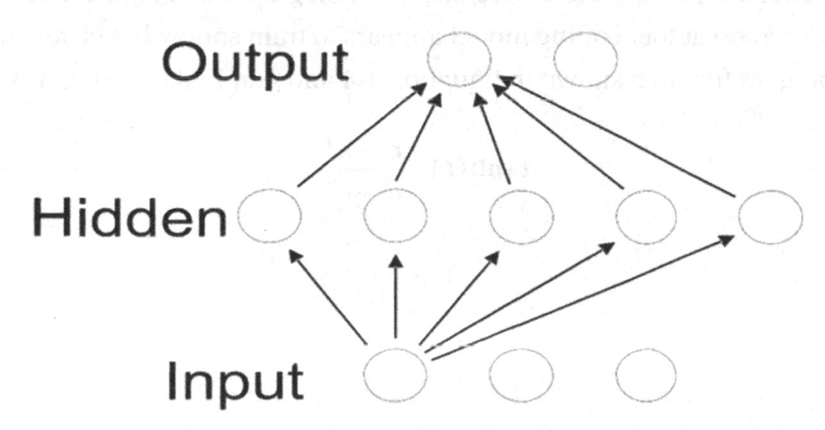

***Figure 4-1.** A feedforward network with one hidden layer*

To be more specific, the activation of a neuron, expressed as a function of the activations of the prior layer, is shown in Equation 4-1. In this equation, $x = \{x_1, \ldots, x_K\}$ is the vector of prior-layer activations, $w = \{w_1, \ldots, w_K\}$ is the vector of associated weights, and b is a bias term.

$$a = f\left(b + \sum_{k=1}^{K} w_k x_k\right) \tag{4-1}$$

It's often more convenient to consider the activation of an entire layer at once. In Equation 4-2, the weight matrix W has K columns, one for each neuron in the prior layer and as many rows as there are neurons in the layer being computed. The bias and layer inputs are column vectors. The nonlinear activation function is applied element-wise to the vector.

$$a = f(b + Wx) \tag{4-2}$$

There is one more way of expressing the computation of activations that is most convenient in some situations. The bias vector b can be a nuisance, so it can be absorbed into the weight matrix W by appending it as one more column at the right side. We then augment the x vector by appending 1.0 to it: $x = \{x_1, \ldots, x_K, 1\}$. The equation for the layer's activations then simplifies to the activation function operating on a simple matrix/vector multiplication, as shown in Equation 4-3.

$$a = f(Wx) \tag{4-3}$$

What about the activation function? The RBM/supervised model and the real-domain autoencoding model use the logistic function graphed in Figure 4-2. However, the complex-domain autoencoding model appears to train somewhat faster with the hyperbolic tangent function shown in Equation 4-4 and graphed in Figure 4-3.

$$\tanh(t) = \frac{e^t - e^{-t}}{e^t + e^{-t}} \tag{4-4}$$

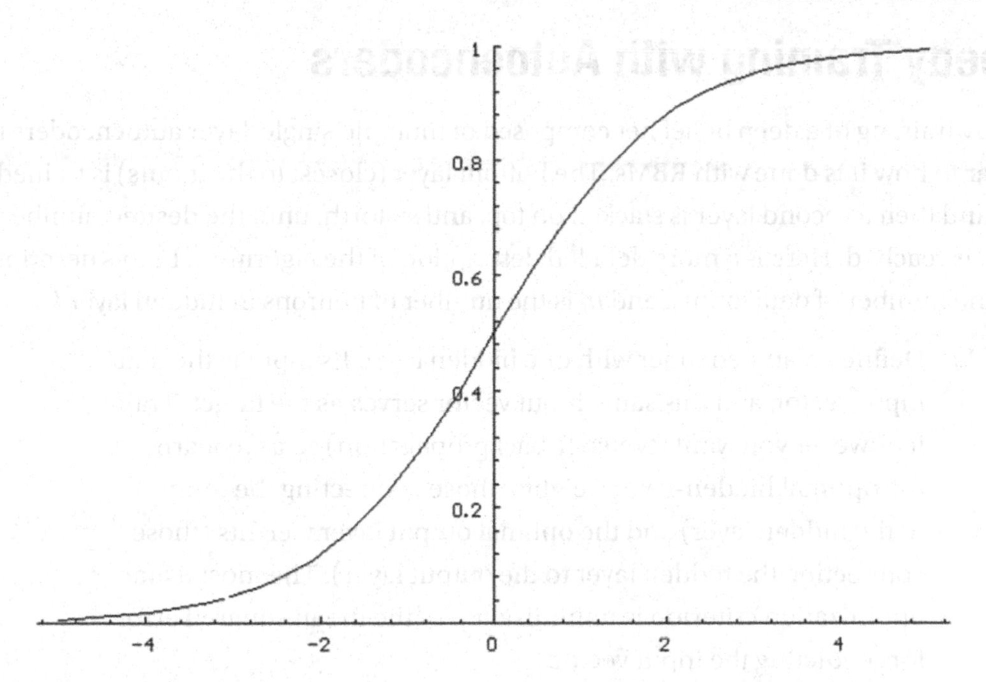

***Figure 4-2.** Logistic function*

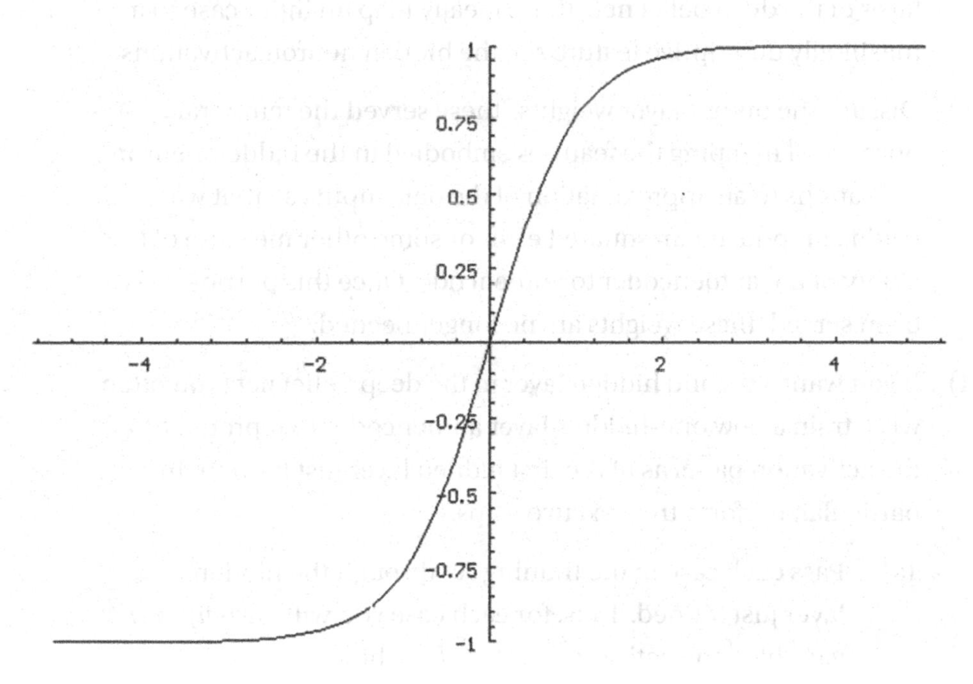

***Figure 4-3.** Hyperbolic tangent function*

Greedy Training with Autoencoders

Greedy training of a deep belief net composed of multiple single-layer autoencoders is similar to how it is done with RBMs. The bottom layer (closest to the inputs) is trained first, and then a second layer is stacked on top, and so forth, until the desired number of layers is reached. Here is a more detailed description of the algorithm. In this description, n_0 is the number of data inputs, and n_k is the number of neurons in hidden layer k.

1) Define an autoencoder with one hidden layer. Its input is the data input vector, and this same input vector serves as the target. Train it however you want (typically backpropagation) so as to learn the optimal hidden-layer weights (those connecting the input to the hidden layer) and the optimal output layer weights (those connecting the hidden layer to the output layer). The most usual optimization criterion is minimization of the mean squared error for predicting the input vector.

2) Save the n_1 hidden-layer weights as the weights for the first hidden layer of the deep belief net. These ideally map an input case to a maximally descriptive feature set, the hidden neuron activations.

3) Discard the output layer weights. These served the temporary purpose of mapping the features embodied in the hidden neuron activations to an approximation of the data inputs so that we could compute mean squared error or some other measure of the ability of the autoencoder to autoencode. Once this purpose has been served, these weights are no longer needed.

4) If you want a second hidden layer in the deep belief net (you often will), train a new one-hidden-layer autoencoder to reproduce the activation patterns of the first hidden layer just trained. In particular, perform the next two steps.

    4a) Pass each case in the training set through the hidden layer just trained. Thus, for each case you will have n_1 new variables, the activation vector of the hidden layer. This collection will serve as the training set for the autoencoder about to be trained. This autoencoder will have n_1 inputs and outputs and n_2 hidden neurons.

4b) Train this new autoencoder. Save the hidden-layer weights as the weights for the second hidden layer of the deep belief net. Again, we discard the output layer weights as their purpose has been served.

5) At this point we have an optional procedure to consider. We have two hidden layers in our deep belief net. Each was trained independently. We can, if we wish, jointly fine-tune these weights to produce a two-hidden-layer autoencoder for the data inputs. In other words, we would train this two-hidden-layer network so that its n_0 outputs will optimally reproduce the n_0 data inputs. Training two hidden layers simultaneously is much more difficult than training each separately. Countering this problem is the fact that the first hidden layer was trained to reproduce the inputs, and the second hidden layer was trained to reproduce the first hidden layer. So, the weights will probably be fairly close to optimal values already. With any luck, convergence will be reasonably fast. But is this fine-tuning a good thing to do? The answer is dependent on the application.

6) If a third hidden layer for the deep belief net is desired, pass each case in the training set through both of the hidden layers just trained. Thus, for each case you will have n_2 new variables, the activation vector of the second hidden layer. This collection will serve as the training set for the autoencoder about to be trained. This autoencoder will have n_2 inputs and outputs and n_3 hidden neurons. Train it, save its hidden-layer weights to serve as the deep belief net's third hidden layer, and discard the output layer weights.

7) Optionally fine-tune these three hidden layers as an autoencoder for the data inputs. This can be a computationally intractable problem or not.

8) Repeat as desired to build the entire deep belief net.

Review of Complex Numbers

This section contains a cursory review of what complex numbers are and how we perform basic arithmetic operations on them. Understanding this material is a prerequisite to understanding most of the remainder of this chapter. No attempt at strict mathematical rigor will be made. The presumed audience is made up of persons having a basic grasp of introductory-level college mathematics but needing a review of concepts important to this chapter.

Complex numbers may be thought of as pairs of real numbers, (*a*, *b*), for which the operations of addition and multiplication are defined in a special way, as shown in Equation 4-5.

$$\begin{aligned}(a,b)+(c,d)&=(a+c,\,b+d)\\(a,b)\cdot(c,d)&=(ac-bd,\,ad+bc)\end{aligned}\tag{4-5}$$

Ambitious readers may want to verify that the previous operations satisfy the field axioms. They are associative and commutative, and the distributive law applies. The additive identity is (0, 0), and the multiplicative identity is (1, 0). Every nonzero complex number has a multiplicative inverse that will be shown as soon as a few other definitions are in place.

It is convenient to write a as shorthand for the complex number $(a, 0)$. This allows us to consider the real numbers to be a subset of the complex numbers. We also define the special quantity $i = (0, 1)$. This means we can write (a, b) as $a + bi$, which is just what we will do from now on.

Applying the rule for multiplication, we see that $i^2 = -1$. The fact that i is the square root of -1 is probably the most famous feature of complex numbers. We are also in a position to define the reciprocal of a nonzero complex number, as shown in Equation 4-5.

$$\frac{1}{a+bi}=\frac{a-bi}{(a+bi)(a-bi)}=\frac{a}{a^2+b^2}+\left(\frac{-b}{a^2+b^2}\right)i\tag{4-6}$$

One special definition with which we should be familiar is the conjugate of a complex number. This is computed by flipping the sign of the imaginary part. It is usually written by placing a bar above the variable, as shown in Equation 4-7.

$$\text{if } z=a+bi \text{ then } \bar{z}=a-bi\tag{4-7}$$

It is often profitable to think of complex numbers as points in the Cartesian plane. This facilitates representing them in polar coordinates. Look at Figure 4-4, which shows the single complex number $a + bi$ plotted as a point.

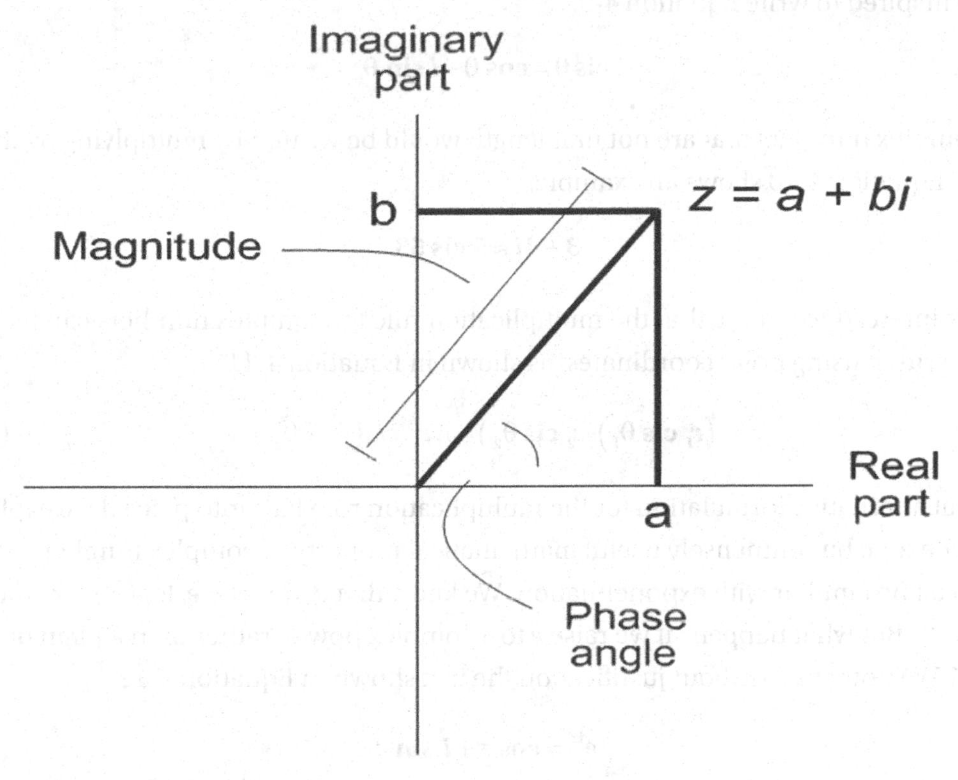

Figure 4-4. *A complex number on a plane*

The absolute value or *magnitude* of a complex number is defined as its length, as shown in Figure 4-4. The *argument* of z, written *arg*(z), is the counterclockwise angle that the vector from the origin to z makes with the positive real axis. These definitions are formalized in Equation 4-8.

$$\begin{aligned} z &= a + bi \\ |z| &= \sqrt{a^2 + b^2} \\ a &= |z| \cos(arg(z)) \\ b &= |z| \sin(arg(z)) \end{aligned} \tag{4-8}$$

The notion of a complex number being expressed in polar coordinates gives us a frequently useful alternative to the $a + bi$ notation. Observing that the real part of a unit length complex number is the cosine of its argument and the imaginary part is the sine, we are inspired to write Equation 4-9.

$$\mathbf{cis}\,\theta = \mathbf{cos}\,\theta + \boldsymbol{i}\,\mathbf{sin}\,\theta \tag{4-9}$$

Complex numbers that are not unit length would be written by multiplying by the length. Equation 4-10 shows an example.

$$\mathbf{3 + 4}\boldsymbol{i} \approx \mathbf{5\ cis\ 53} \tag{4-10}$$

It is interesting to note that the multiplication rule for complex numbers can be nicely written using polar coordinates, as shown in Equation 4-11.

$$(r_1\,\mathbf{cis}\,\theta_1)(r_2\,\mathbf{cis}\,\theta_2) = r_1 r_2\ \boldsymbol{cis}(\theta_1 + \theta_2) \tag{4-11}$$

That alternative formulation for the multiplication rule falls into place if we explore a rather exotic but immensely useful mathematical property of complex numbers. We should all be familiar with exponentiation. We know that $e^0 = 1$, $e^1 = e$, $\log(e^x) = x$, and $e^a e^b = e^{a+b}$. But what happens if we raise e to a complex power, rather than a plain old real power? We now state without justification the fact shown in Equation 4-12.

$$e^{iz} = \mathbf{cos}\,z + i\,\mathbf{sin}\,z \tag{4-12}$$

Although Equation 4-12 is true for any complex value of z, we will here be concerned with strictly real values of z. You should carefully ponder how that equation gives us an alternative form of polar representation because that may help some people visualize derivatives in the complex domain. And how about that multiplication rule now?

Later in this chapter we will be discussing functions that map complex numbers to other complex numbers. In that context, we will also need to consider derivatives of these functions. Let's now briefly review derivatives in the real domain, hint at what fierce things derivatives are in the complex domain, and finally show how we will sidestep that perilous field by working with partial derivatives.

You may recall from basic calculus that the derivative of a function (from the reals to the reals) is the relative degree to which that function changes in response to a small change in its domain variable, as shown in Equation 4-13.

$$f'(x) = \lim_{h \to 0} \frac{f(x+h) - f(x)}{h} \tag{4-13}$$

Implicit in that definition are that the limit exists and that it is the same regardless of whether we approach x from the left (negative h) or from the right (positive h).

The fact that we can approach x from only the left or the right makes real-domain derivatives straightforward. Would that it were so simple in the complex domain. Here, there are an infinite number of directions from which we can approach a point. And the limit of the relative change in the function value must be the same for all of those directions! Little thought is needed to become convinced that this is a very stiff requirement. If we have some function that has a derivative throughout its domain and if that derivative is continuous, then we say that our function is analytic. This term will come up later, so it was roughly defined here. But this is as deep as we need to dig.

We will often be dealing with functions whose domain is real-valued vectors and whose range is the reals. Such functions can have partial derivatives with respect to any element of the vector. (Higher-order derivatives will not be discussed.) This is the relative change effected in the function value in response to a small change in one of the components of the domain vector. Equation 4-14 shows an example.

$$\frac{\partial}{\partial x} f(x,y) = \lim_{h \to 0} \frac{f(x+h,y) - f(x,y)}{h} \tag{4-14}$$

When we have a function from the complex domain to the complex domain, derivatives become much simpler if we separate the real and imaginary parts of the domain and range. Treat it as two functions, each from the complex domain to the real range. One of these functions defines the real part of the complex-valued function, and the other defines the imaginary part.

$$f(a+bi) = f_r(a+bi) + f_i(a+bi)\,i \tag{4-15}$$

Then we can deal with four partial derivatives: the real part of the function with respect to the real part of the domain variable, the imaginary part of the function with respect to the real part of the domain variable, the real part of the function with respect to the imaginary part of the domain variable, and the imaginary part of the function with respect to the imaginary part of the domain variable. Functions from the complex domain to the complex domain for which these four partial derivatives exist and are continuous are vastly more common than functions that are analytic.

Fast Dot Product Computation in the Complex Domain

When training a neural network, the primary eater of time is computation of dot products. It is well known (or at least fairly well known!) that loop unrolling can often significantly speed dot product computation. We saw the basic rules for adding and multiplying complex numbers in Equation 4-5. So now we put together these two concepts for a routine that computes the dot product of two complex vectors. This code is listed on the next page. As will be universal throughout most of this text, complex vectors are stored as pairs, with the real part followed by the imaginary part.

If you are unfamiliar with loop unrolling, the idea is to handle vector computation in chunks. The most obvious benefit is that checking the value of the loop index and usually incrementing it and jumping back to the start of the loop happens less frequently. But an even greater benefit is that most modern processors can look ahead and perform memory fetches and unrelated computations in parallel. It is always in our best interest to write programs that make this as easy as possible for the processor. Finally, by keeping nearby in code those variables that are reused, we encourage the compiler to store them in registers.

```
void dotprodc (
  int n ,             // Length of vectors
  double *vec1 , // One of the vectors to be dotted
  double *vec2 , // The other vector
  double *re ,      // Real part of output
  double *im )     // and imaginary
{
  int k, m ;

  *re = *im = 0.0 ;                // Will cumulate dot product here
  k = n / 4 ;                        // Divide vector into this many groups of 4
  m = n % 4 ;                       // This is the remainder of that division

  while (k--) {                     // Do each group of 4
    *re += *(vec1 ) * *(vec2 ) - *(vec1+1) * *(vec2+1) ;
    *im += *(vec1 ) * *(vec2+1) + *(vec1+1) * *(vec2 ) ;
    *re += *(vec1+2) * *(vec2+2) - *(vec1+3) * *(vec2+3) ;
    *im += *(vec1+2) * *(vec2+3) + *(vec1+3) * *(vec2+2) ;
    *re += *(vec1+4) * *(vec2+4) - *(vec1+5) * *(vec2+5) ;
```

```
      *im += *(vec1+4) * *(vec2+5) + *(vec1+5) * *(vec2+4) ;
      *re += *(vec1+6) * *(vec2+6) - *(vec1+7) * *(vec2+7) ;
      *im += *(vec1+6) * *(vec2+7) + *(vec1+7) * *(vec2+6) ;
      vec1 += 8 ;
      vec2 += 8 ;
      }

   while (m--) {               // Do the remainder
      *re += *vec1 * *vec2 - *(vec1+1) * *(vec2+1) ;
      *im += *vec1 * *(vec2+1) + *(vec1+1) * *vec2 ;
      vec1 += 2 ;
      vec2 += 2 ;
      }
}
```

Singular Value Decomposition in the Complex Domain

You should be familiar with linear regression. We presuppose that a dataset can be explained by an underlying linear model, plus a bit of noise contamination. The expected value for a target variable y is given by a weighted sum of predictors a_i plus a constant. Equation 4-16 is a typical statement of the underlying model. In this equation we have a total of m values, w_0 through w_{m-1}, that define the model. The first m-1 of these weights (subscripts 0 through m-2) multiply the predictors, and the last is the constant, often called the bias term.

$$\hat{y} = a_0 w_0 + a_1 w_1 + \ldots + w_{m-1} \tag{4-16}$$

The "traditional" method of solving for optimal values for the weights involves inversion of a matrix that, in many practical applications, will be singular or nearly so. To avoid this deadly problem, conscientious developers employ the technique of singular value decomposition.

A subroutine for performing singular value decomposition, SVDCMP.CPP, is available for free download from my web site. This file contains detailed instructions for its use in solving linear regression problems, so we'll provide just an overview here. The basic idea is that first the user fills in a matrix, each of whose rows contains the m-1 values of the predictors for a case, followed by 1.0 to represent the bias constant. The member function svdcmp() is called. Then the user fills in a vector containing the target values corresponding to the cases just entered, and the member function backsub() is called. This returns the weights that minimize the mean squared error of the predictions.

How does this relate to training neural networks? Recall that the output layer is linear. Thus, if we compute the activations of the layer just prior to the output layer, we can use these as predictors in linear regression and thereby compute optimal weights for the output layer, without having to employ an expensive iterative process. This is worth its weight in, well, whatever is most valuable to you.

The method just described is easily extended to the case of complex-valued predictors and targets. All we do is treat the real and imaginary parts separately. Each case in the training set now generates two rows in the *a* matrix and two entries in the target vector. One of these rows is for predicting the real part of the target, and the other is for predicting the imaginary part. For a given case, these two equations would be written as shown next. If necessary, look back at Equation 4-5 to see how the pairs of terms come about. Figure 4-5 illustrates the layout of the data in the *a* matrix and *y* vector.

$$\begin{aligned} a_{0,r} w_{0,r} - a_{0,i} w_{0,i} + \dots \quad + w_{m-1,r} &= y_r \\ a_{0,i} w_{0,r} + a_{0,r} w_{0,i} + \dots \quad + w_{m-1,i} &= y_i \end{aligned} \tag{4-17}$$

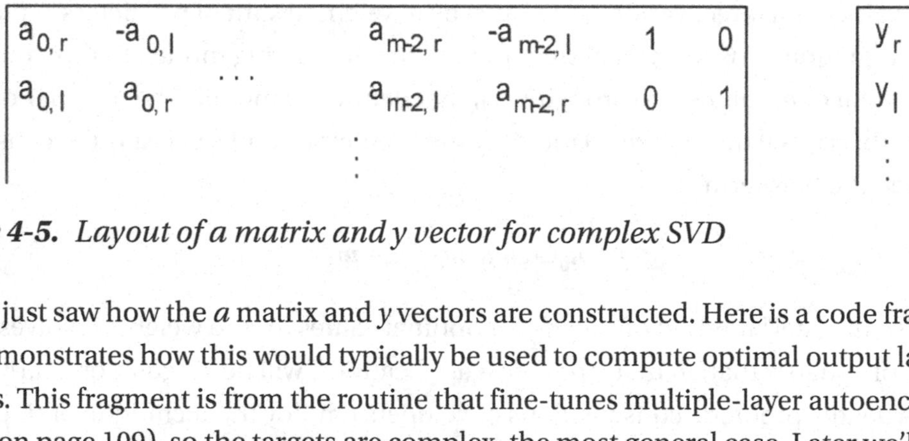

$$\begin{bmatrix} a_{0,r} & -a_{0,I} & \cdots & a_{m-2,r} & -a_{m-2,I} & 1 & 0 \\ a_{0,I} & a_{0,r} & & a_{m-2,I} & a_{m-2,r} & 0 & 1 \\ & & & \vdots & & & \end{bmatrix} \qquad \begin{bmatrix} y_r \\ y_I \\ \vdots \end{bmatrix}$$

Figure 4-5. *Layout of a matrix and y vector for complex SVD*

We just saw how the *a* matrix and *y* vectors are constructed. Here is a code fragment that demonstrates how this would typically be used to compute optimal output layer weights. This fragment is from the routine that fine-tunes multiple-layer autoencoders (step 5 on page 109), so the targets are complex, the most general case. Later we'll see that for fine-tuning the complete network we consider only the real parts of the targets, and we'll see that the algorithm used there is a subset of the algorithm shown here.

We create the SingularValueDecomp object that does the work. The first parameter is the number of rows in the *a* matrix. In the full complex application, there are two rows for each case. The second parameter is the number of columns in the matrix. There are nhid hidden neurons whose activations are feeding the output layer, along with a bias constant.

Each of these quantities is full complex and hence occupies two columns. The last parameter tells the constructor that it does not need to preserve the *a* matrix for later use.

```
svd = new SingularValueDecomp ( 2 * n_cases , 2 * (nhid + 1) , 0 ) ;
```

We fetch the input data and propagate it through the hidden layer(s). The activations of the final hidden layer, that feeding the output layer, are stored in the matrix `tmp_inputs`, which has a column dimension of `max_neurons`. In general, this column dimension may exceed the number of columns actually used. The internal operations of the input-fetching and propagation routines are not shown because they are dependent on the architecture of the program; your implementation would surely be different from mine, and moreover they have nothing to do with the topic at hand.

```
data_to_tmp () ;                      // Get the input data
propagate ( n_cases , n_layers ) ;    // And propagate through layers-so-far
```

We now build half of the *a* matrix and *y* vector, those rows corresponding to the real part of the targets.

```
aptr = svd->a ;                              // The a matrix goes here

for (i=0 ; i<n_cases ; i++) {                // These rows are for the real predictions
  dptr = tmp_inputs + i * max_neurons ;      // Point to this case in the activations
  for (j=0 ; j<nhid ; j++) {                 // Activation of each hidden neuron
    *aptr++ = *dptr++ ;                      // Real part of activation
    *aptr++ = - *dptr++ ;                    // And imaginary part
    }
  *aptr++ = 1.0 ;                            // Constant for real bias
  *aptr++ = 0.0 ;                            // Rows are real, so no Imaginary bias
  }
```

Now we build the rows that correspond to predicting the imaginary component of the targets.

```
for (i=0 ; i<n_cases ; i++) {                // This set is for the imaginary predictions
  dptr = tmp_inputs + i * max_neurons ;      // Point to this case in the activations
  for (j=0 ; j<nhid ; j++) {                 // Activation of each hidden neuron
    *aptr++ = dptr[1] ;                      // Imaginary part of activation
```

```
      *aptr++ = dptr[0] ;                       // And real part
      dptr += 2 ;                               // Advance pointer to next neuron
      }
    *aptr++ = 0.0 ;                             // Predicting imaginary, so no real bias
    *aptr++ = 1.0 ;                             // Bias for imaginary part of target
    }
```

Perform the singular value decomposition. Once again we fetch the input data, but we don't need to propagate it through the hidden layer(s). All we need the inputs for in this particular application is to serve as targets for what will be an autoencoder. If your network is being trained for a more general purpose, then you'll fetch whatever targets you want.

```
  svd->svdcmp () ;
  data_to_tmp () ;                            // Get the targets
```

Each target is handled separately. We find the weights that connect the final hidden layer (whose activations are in the *a* matrix) to a single output neuron, one output at a time.

```
  for (i=0 ; i<n_targets ; i++) {             // For each target (output neuron)
    bptr = svd->b ;                           // The y vector must be placed here

    for (j=0 ; j<n_cases ; j++) {             // Handle the real parts of the target
      dptr = tmp_inputs + j * max_neurons ;   // Point to target case
      *bptr++ = dptr[2*i] ;                   // Real part of target
      }
    for (j=0 ; j<n_cases ; j++) {             // Handle the imaginary parts of the target
      dptr = tmp_inputs + j * max_neurons ;   // Point to target case
      *bptr++ = dptr[2*i+1] ;                 // Imaginary part of target
      }

    svd->backsub ( 1.e-4 , output_weights + i * 2 * (nhid + 1) ) ; // Find output weights
    }
```

Activation in the Complex Domain

The section beginning on page 105 presented an overview of feedforward neural networks. It was pointed out that the hyperbolic tangent function shown in Equation 4-4 will form the foundation of activation in the complex domain. We now explore this issue.

That discussion was confined to data, targets, and activation functions in the real domain. It is now time to take our first step into the complex domain. Things are not so neat and tidy here. One of the most complicating factors is that derivatives are not straightforward extensions of their real-domain counterparts. A real point can be approached only from the left and right. A point in the complex domain can be approached from any direction in the complex plane. This makes even the very existence of derivatives in the complex domain a much more serious affair. We will sidestep this problem by avoiding direct use of complex derivatives. Rather, we will write all complex numbers as real and imaginary parts and refer to the partial derivatives of the real and imaginary parts of the function with respect to the real and imaginary parts of its domain variable. As will be seen later, this gives us far more latitude in choosing an activation function.

It is not generally possible to find complex extensions of real activation functions that are usable in the complex domain. For example, it is well-known that the hyperbolic tangent function is an excellent real-domain activation function. Now take a look at the real and imaginary parts of the *tanh* function in the complex domain. These are plotted (with vertical truncation) in Figures 4-6 and 4-7, respectively. Common sense tells us in no uncertain terms that this does not look like a useful activation function! Moreover, the fact that it has violent discontinuities across its domain precludes its use if gradients are to be computed for learning algorithms. Something better is needed.

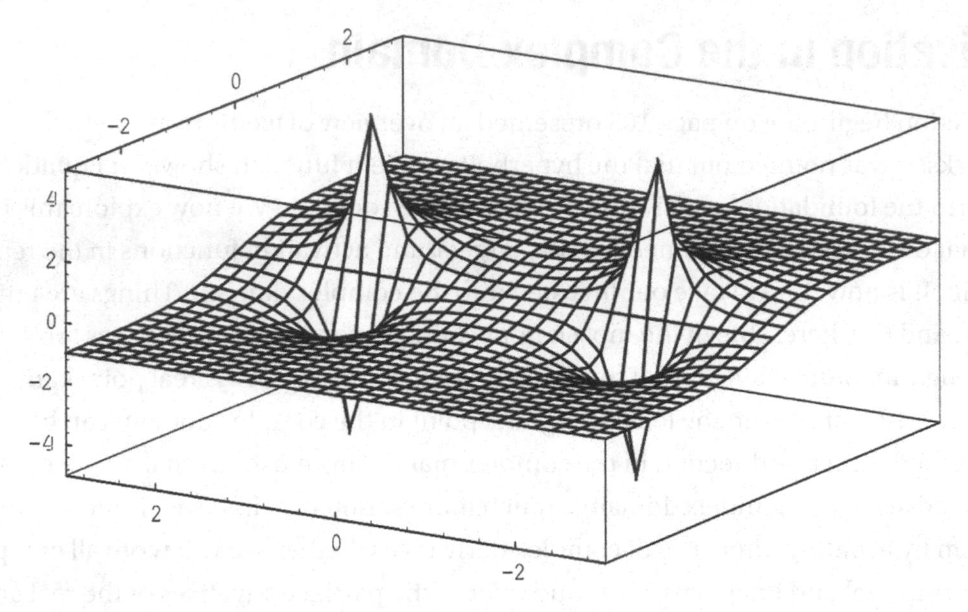

Figure 4-6. *Real part of complex tanh function*

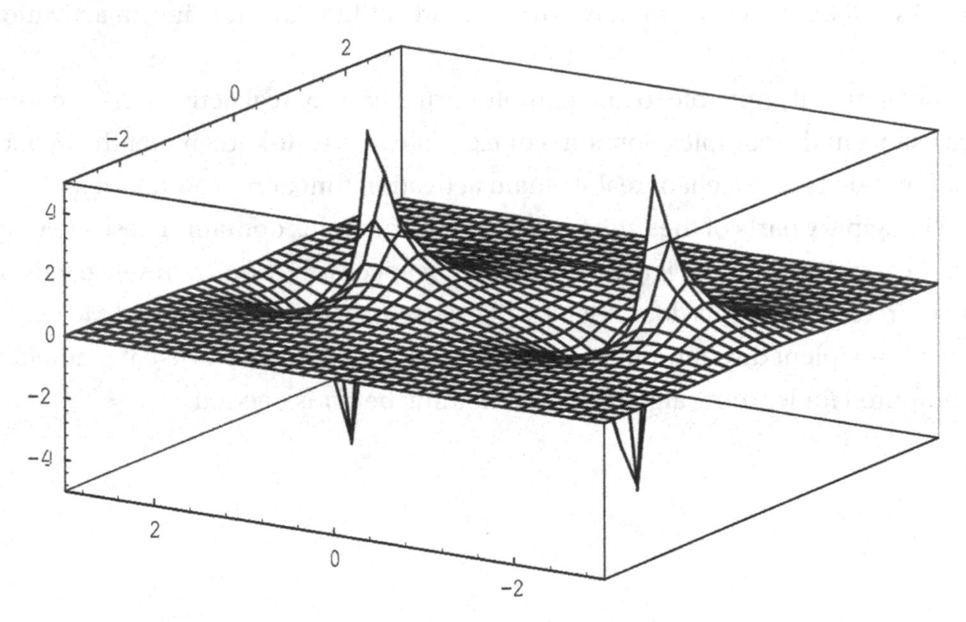

Figure 4-7. *Imaginary part of complex tanh function*

Many years of experience in the real domain has led to widespread agreement that the most effective activation functions have a *sigmoid* (S-shaped) form. A sigmoid is a function from the reals to the reals that is bounded, differentiable, and has a positive derivative everywhere. By convention, sigmoid activation functions are also assumed to rapidly approach their bounds asymptotically and to be roughly linear near the center of their domain (typically zero).

Now consider the properties that a good activation function in the complex domain should have. These properties are intuitive extensions of the definition of a sigmoid function in the real domain. Thus, we would certainly like its real and imaginary parts to be continuously differentiable with respect to the real and imaginary parts of the domain variable. (It is unreasonable to expect the function to be analytic, though. This would impose a severe and unnecessary restriction on our choices.) We would like its magnitude to be bounded, and we would like it to approach these bounds rapidly as the magnitude of the domain variable increases. Finally, we would like the function to be approximately linear when the magnitude of the domain variable is small.

An immediately obvious possibility is to apply some squashing function to the real and imaginary parts separately. Equation 4-18 shows an example.

$$f(x+yi)=\tanh(x)+\tanh(y)i \tag{4-18}$$

Some researchers have reported good results using this technique.

However, there is another appealing property that functions like that shown in Equation 4-18 do not have. It would be nice if the activation function preserved the direction of the net input, squashing only its length. This is easily implemented. Suppose that $s(x)$ is a sigmoid function having $s(0) = 0$. The hyperbolic tangent is one such function. We will be concerned only with non-negative values of its domain variable. Then we can squash the length of a complex number without changing its direction by means of the function shown in Equation 4-19.

$$\begin{gathered} f(x+yi)=px+pyi \\ p=\frac{s\left(\sqrt{x^2+y^2}\right)}{\sqrt{x^2+y^2}} \end{gathered} \tag{4-19}$$

The real and imaginary parts of that function for $s(x) = \tanh(x)$ are graphed in Figures 4-8 and 4-9, respectively. That activation function computes the output by simply multiplying the net input by a factor. The factor is the ratio by which the real-domain squashing function, $s(x)$, compresses the length of the net input. The activation function defined by Equation 4-19 when $s(x) = \tanh(1.5\ x)$ is my standard workhorse and will be used throughout this text. The factor of 1.5 is suggested by some researchers but is certainly not vital.

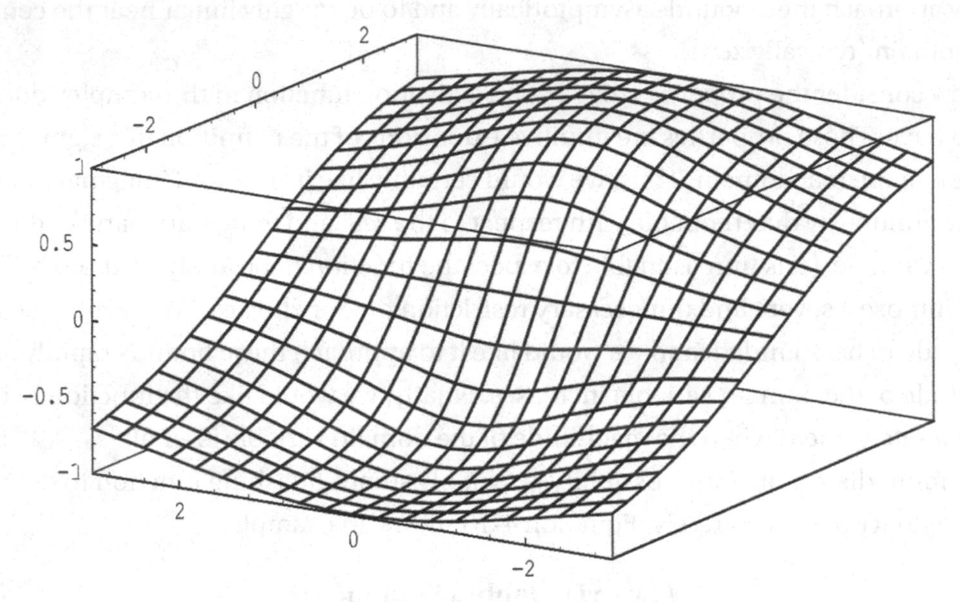

Figure 4-8. *Real part of complex activation function*

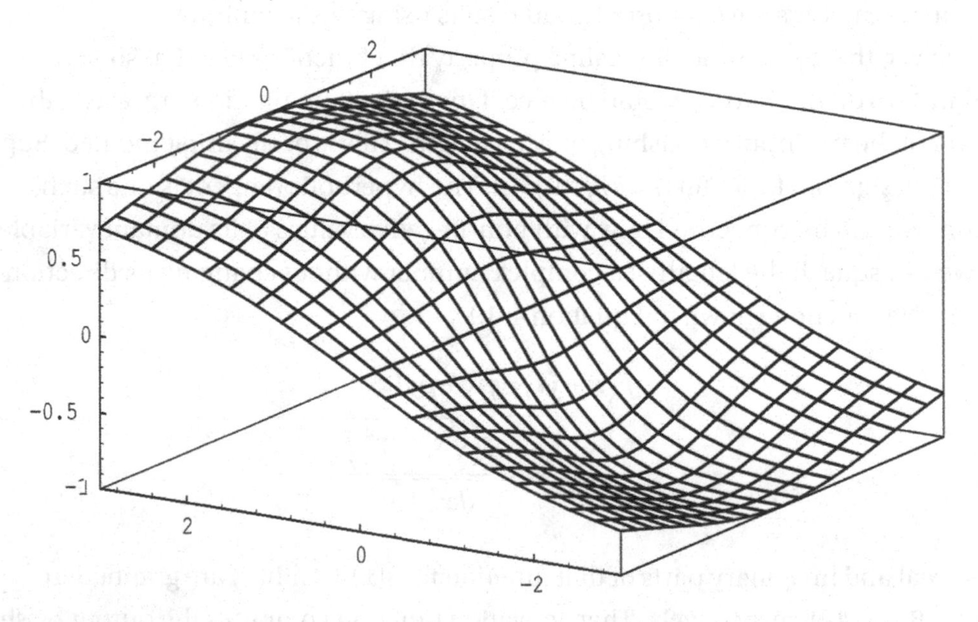

Figure 4-9. *Imaginary part of complex activation function*

Derivatives of the Activation Function

In the next section, when we compute the error gradient, we will need the derivatives of the activation function. This section will show how to compute these derivatives for the activation function expressed by Equation 4-19.

Let us start by rewriting half of Equation 4-19 as a function of two real variables. It should be apparent that we can use Equation 4-20 to compute both the real and imaginary parts of the activation function according to whether we let x take the role of the real or imaginary part of the net input, respectively.

$$h(x,y)=\frac{xs\left(\sqrt{x^2+y^2}\right)}{\sqrt{x^2+y^2}} \tag{4-20}$$

To simplify the next few equations, define the length of the net input.

$$L=\sqrt{x^2+y^2} \tag{4-21}$$

We can now write the partial derivative of $h(x,y)$ with respect to x.

$$\frac{\partial h}{\partial x}=\frac{s(L)}{L}-\frac{x^2 s(L)}{L^3}+\frac{x^2 s'(L)}{L^2} \tag{4-22}$$

When Equation 4-20 is used to compute the real part of the activation function by letting x play the role of the real part of the net input (with y being the imaginary part), Equation 4-22 gives us the partial derivative of the real part of the output with respect to the real part of the input. Similarly, when x is the imaginary part of the input and y is the real part, Equation 4-22 is the partial derivative of the imaginary part of the output with respect to the imaginary part of the input.

To complete the picture, we need the partial derivative of $h(x,y)$ with respect to y, as shown in Equation 4-23.

$$\frac{\partial h}{\partial x}=\frac{xys'(L)}{L^2}-\frac{xys(L)}{L^3} \tag{4-23}$$

This equation allows us to compute the partial derivative of the real part of the output with respect to the imaginary part of the input, and the partial derivative of the imaginary part of the output with respect to the real part of the input. Because of the obvious symmetry, these two quantities are equal.

The derivative of our squashing function is shown in Equation 4-24.

$$\begin{aligned} s(x) &= \tanh(1.5x) \\ s'(x) &= 1.5\left(1 - s^2(x)\right) \end{aligned} \tag{4-24}$$

Code for computing the partial derivatives is shown next. Several features of this code should be noted. First, as is frequently the case, it is most efficient to compute both the activation and its partial derivatives simultaneously, so that is what we do. Second, note the provision for a linear activation function. If this is an output neuron, we usually want linearity. In this case, computations are trivial.

The input and coefficient vectors contain complex numbers stored as pairs. The real part of each element is immediately followed by the imaginary part. The output is a single complex number and is stored the same way. The calling parameter list is as follows:

```
void activity_cc (
  double *input ,      // This neuron's input vector, 2 * ninputs long
  double *coefs ,      // Weight vector, 2 * (ninputs+1) long (bias is at end)
  double *output ,     // Achieved activation of this neuron (real, imag)
  double *d_rr ,       // If non-null, returns partial of real activation wrt real input
  double *d_ii ,       // Ditto, imag wrt imag
  double *d_ri ,       // Ditto, real wrt imag, which equals imag wrt real
  int ninputs ,        // Number of complex inputs
  int outlin           // Activation function is identity if nonzero, else logistic
  )
```

First we compute the complex-domain dot product of the inputs with the weights, using the routine shown on page 114. Then we add in the bias. If this is linear, we just return the activations. The derivatives of a linear activation are simply 1.0 for the real/real and imaginary/imaginary, and 0.0 for the crossovers, so it is pointless to compute them. It's easier for the caller to code them in.

```
{
  double rsum, isum, raw_length, squashed_length, ratio, deriv, len_sq, temp ;

  dotprodc ( ninputs , input , coefs , &rsum , &isum ) ;
  rsum += coefs[2*ninputs] ; // Bias term
  isum += coefs[2*ninputs+1] ;
```

```
   if (outlin) {
      *output = rsum ;
      *(output+1) = isum ;
      return ;
      }
```

We use Equation 4-20 to compute the real and imaginary parts of the activation. The offset of 1.e-60 is to prevent division by zero later in the code.

```
   len_sq = rsum * rsum + isum * isum + 1.e-60 ;
   raw_length = sqrt ( len_sq ) ;
   squashed_length = tanh ( 1.5 * raw_length ) ;      // 1.5 is heuristic but good
   ratio = squashed_length / raw_length ;              // This is p in Equation 4-19

   *output = rsum * ratio ;
   *(output+1) = isum * ratio ;
```

Sometimes the caller will want only the activation function, and sometimes it will also want the partial derivatives.

```
   if (d_rr == NULL)
      return ;
```

The last few lines implement Equations 4-22 and 4-23. The relation between these equations and the code may not be immediately obvious because the code has been designed to be fast and accurate.

```
   deriv = 1.5 * (1.0 - squashed_length * squashed_length) ; // Equation 4-24
   temp = (deriv - ratio) / len_sq ;

   *d_rr = ratio + rsum * rsum * temp ;
   *d_ii = ratio + isum * isum * temp ;
   *d_ri = rsum * isum * temp ;
}
```

The Logistic Activation Function and Its Derivative

We are primarily concerned with complex-domain models here. But the DEEP program also allows real-domain models, and in this case it uses the logistic activation function. It and its derivative are shown in Equation 4-25.

$$
\begin{aligned}
f(x) &= \frac{1}{\left(1+e^{-x}\right)} \\
f'(x) &= f(x)\left(1-f(x)\right)
\end{aligned}
\tag{4-25}
$$

Computing the Gradient

Many standard neural network textbooks derive the equations for computing the gradient of the mean squared error for real-domain networks. Volume I of this series has a detailed treatment of the topic. You will probably benefit from taking time to review this subject to be clear on the process of backpropagation of errors for computing the gradient.

Because there is an infinite number of directions from which a point in the complex plane can be approached, derivatives in the complex domain are not always a straightforward extension of their real-domain counterparts. It is, therefore, worthwhile to dedicate several tedious pages to explicit derivation of the gradient for complex-valued neural networks.

Several special notation conventions will be employed here. Their purpose is to minimize complexity in what could otherwise be an overwhelming exhibition. Any ambiguities will be easily resolved from context.

Our goal in the following is to compute the partial derivative of a three-layer network's error with respect to a single weight connecting a neuron in one layer to a neuron in another layer. Extensions to more hidden layers are easily done by recursion, and this will be illustrated with program code. The gradient will be found for a single input presentation. The gradient for an entire training epoch is the sum of the gradients for all presentations in that epoch.

Traditional real-domain derivations use subscripts on the weight, w_{ij}, to indicate the source and destination neurons. We will avoid this because the subscripts introduce unnecessary complexity. Instead, we will use a subscript of "r" to indicate that we are referring to the real part of the weight, and a subscript of "i" for the imaginary part.

Thus, a single complex weight would be $w = (w_r + w_i i)$. It will be seen that subscripts indicating the neurons being connected by the weight are nearly always superfluous. They will be used only when absolutely necessary.

We now give definitions of the terms that will appear. The error, E, is a real number. All other terms are complex. Therefore, like the weight described earlier, they may have a subscript of "r" or "i" to indicate whether they refer to the real or imaginary part. The subscript is not shown in the definitions here.

in: A single input to the network

hnet: The net input to a hidden neuron

h: The activation of that hidden neuron

anet: The net input to an output neuron

a: The activation of that output neuron

t: The target activation for that output neuron

E: The error for that output neuron

E_{TOT}: The total error for all output neurons

w: A weight connecting one neuron to another

E refers to the error of the single output neuron under consideration when it does not have a subscript. Occasionally, though, it will appear as part of a summation across all output neurons. In these cases, it will be subscripted to indicate that it refers to a particular neuron. So, for example, we have Equation 4-26.

$$\boldsymbol{E}_{\mathbf{TOT}} = \sum_k \boldsymbol{E}_k \tag{4-26}$$

Note that it is possible to define more complicated network errors than the sum of individual output neuron errors. In particular, *SoftMax* output error for classification will be discussed on page 130.

Unless otherwise noted, all computations will be done with real numbers, using the definitions of complex arithmetic (for example, Equation 4-5) to operate on the real and imaginary parts of complex numbers separately. This is because most of the following equations will be programmed directly, so a strictly real presentation simplifies the translation from equations to code when the target language does not directly support efficient complex arithmetic.

The effect of a weight on the network error is the result of percolation through one or more net computations, one or more activation functions, and a final error function. We will apply tedious repetition of the generalized chain rule, shown in Equation 4-27, to break up that string of operations.

$$\begin{aligned}
\boldsymbol{u} &= \boldsymbol{g}_1(\boldsymbol{x}, \boldsymbol{y}) \\
\boldsymbol{v} &= \boldsymbol{g}_2(\boldsymbol{x}, \boldsymbol{y}) \\
\boldsymbol{h} &= \boldsymbol{f}(\boldsymbol{u}, \boldsymbol{v}) \\
\frac{\partial \boldsymbol{h}}{\partial \boldsymbol{x}} &= \frac{\partial \boldsymbol{h}}{\partial \boldsymbol{u}}\frac{\partial \boldsymbol{u}}{\partial \boldsymbol{x}} + \frac{\partial \boldsymbol{h}}{\partial \boldsymbol{v}}\frac{\partial \boldsymbol{v}}{\partial \boldsymbol{x}} \\
\frac{\partial \boldsymbol{h}}{\partial \boldsymbol{y}} &= \frac{\partial \boldsymbol{h}}{\partial \boldsymbol{u}}\frac{\partial \boldsymbol{u}}{\partial \boldsymbol{y}} + \frac{\partial \boldsymbol{h}}{\partial \boldsymbol{v}}\frac{\partial \boldsymbol{v}}{\partial \boldsymbol{y}}
\end{aligned} \tag{4-27}$$

Let's start with the easy weights, those that connect the hidden layer to the output layer. In particular, we are interested in a weight, $w = (w_r + w_i i)$, connecting a hidden neuron to an output neuron. The activation of the hidden neuron is $h = (h_r + h_i i)$. The net input to the output neuron is *anet*, and the activation of that neuron is *a*. (The breakdown into real and imaginary parts of such terms no longer will be explicitly shown, as you certainly get the idea by now. Also, this text always employs linear outputs so that *a* = *anet*. But for maximum generality, we will not make this assumption in the development of equations.)

Examine once again Equation 4-1, which shows how the activation of a neuron is computed. For now, also assume that the sum of squared errors is used. We will show on page 130 how SoftMax output error for classification can be incorporated into gradient calculations. The error of the single neuron under consideration, which we will call output neuron *k*, is the squared length of the difference between its target and achieved activations, as shown in Equation 4-28.

$$\boldsymbol{E}_k = |\boldsymbol{a} - \boldsymbol{t}|^2 = (\boldsymbol{a}_r - \boldsymbol{t}_r)^2 + (\boldsymbol{a}_i - \boldsymbol{t}_i)^2 \tag{4-28}$$

We begin by using the chain rule to break down our desired partial derivative. Note that we need not be concerned about output neurons other than the one under consideration, as the weight *w* affects only that one. In other words, the partial derivative of the total error with respect to *w* is equal to the partial derivative of the error of that single output neuron with respect to *w*, as shown in Equation 4-29.

$$\frac{\partial E_{TOT}}{\partial w_r} = \frac{\partial E_{TOT}}{\partial anet_r}\frac{\partial anet_r}{\partial w_r} + \frac{\partial E_{TOT}}{\partial anet_i}\frac{\partial anet_i}{\partial w_r}$$
$$\frac{\partial E_{TOT}}{\partial w_i} = \frac{\partial E_{TOT}}{\partial anet_r}\frac{\partial anet_r}{\partial w_i} + \frac{\partial E_{TOT}}{\partial anet_i}\frac{\partial anet_i}{\partial w_i} \tag{4-29}$$

The right-hand factors of the terms in Equation 4-29 are not difficult. Equation 4-1 shows that the net input to this output neuron depends on this weight only through the activation of the single hidden neuron from which this weight leads. Remember how the product of the weight with the hidden neuron activation is computed, as shown in Equation 4-30. (This was Equation 4-5.)

$$\begin{aligned} hw &= (h_r + h_i i)(w_r + w_i i) \\ &= (h_r w_r - h_i w_i) + (h_r w_i + h_i w_r)\, i \end{aligned} \tag{4-30}$$

Thus, we immediately know the right-hand factors in Equation 4-29. They are shown in Equation 4-31.

The left-hand factors in Equation 4-29 require a little more work. The chain rule must be applied again, as shown in Equation 4-32. Note that these factors are labeled δ_{or} and δ_{oi} in that equation. This stands for the real and imaginary parts of the output delta. There are two reasons for this extra label. One is that we will refer to these quantities later in this development, and this is convenient shorthand for an otherwise complicated expression. The other is that most traditional derivations of the gradient for real-domain neurons use δ as an intermediate term at this same point. Retaining this common convention enables you to see the correspondence between the derivation in other standard texts and the derivation here.

$$\frac{\partial anet_r}{\partial w_r} = h_r \qquad \frac{\partial anet_r}{\partial w_i} = -h_i$$
$$\frac{\partial anet_i}{\partial w_r} = h_i \qquad \frac{\partial anet_i}{\partial w_i} = h_r \tag{4-31}$$

$$\delta_{or} = \frac{\partial E_{TOT}}{\partial anet_r} = \frac{\partial E_{TOT}}{\partial a_r}\frac{\partial a_r}{\partial anet_r} + \frac{\partial E_{TOT}}{\partial a_i}\frac{\partial a_i}{\partial anet_r}$$
$$\delta_{oi} = \frac{\partial E_{TOT}}{\partial anet_i} = \frac{\partial E_{TOT}}{\partial a_r}\frac{\partial a_r}{\partial anet_i} + \frac{\partial E_{TOT}}{\partial a_i}\frac{\partial a_i}{\partial anet_i} \tag{4-32}$$

The right-hand factors in Equation 4-32 are the partial derivatives of the output neuron's activation function. In this text and the DEEP program, the output activation is linear, meaning that the real/real and imaginary/imaginary partial derivatives are 1.0, and the real/imaginary and the imaginary/real are 0.0. But for optimal generality, we leave it this way in case you want to program a nonlinear output activation function. Computation of partial derivatives of activation functions was discussed on page 123.

The left-hand factors in Equation 4-32 are dependent on the way we measure the network's error. Most commonly, we assume the traditional sum of squared errors as defined in Equations 4-26 and 4-28. This gives us a relatively simple partial derivative.

$$\begin{aligned} \frac{\partial \boldsymbol{E}_{\mathbf{TOT}}}{\partial \boldsymbol{a}_{\mathrm{r}}} &= \mathbf{2}\left(\boldsymbol{a}_{\mathrm{r}} - \boldsymbol{t}_{\mathrm{r}}\right) \\ \frac{\partial \boldsymbol{E}_{\mathbf{TOT}}}{\partial \boldsymbol{a}_{\mathrm{i}}} &= \mathbf{2}\left(\boldsymbol{a}_{\mathrm{i}} - \boldsymbol{t}_{\mathrm{i}}\right) \end{aligned} \tag{4-33}$$

Pure Real and SoftMax Output Errors

There are a multitude of possible error measures, but here we consider only two other possibilities. Equation 4-33 is appropriate when we are predicting full complex numbers. This would be the case, for example, if we are training a complex-domain autoencoder. But if we have constructed a deep belief net of stacked autoencoders and are now training it to predict a strictly real target, then we use only the top equation in that pair. The partial derivative of the error with respect to the imaginary part is zero because we ignore the imaginary part when computing the error.

Volume I presented an extensive discussion of SoftMax output for classification. You are encouraged to see that discussion for motivation and further details. However, for the sake of those who choose to avoid that, some key points are reproduced here.

We again assume that the bias terms are absorbed into the weights by appending (1.0 + 0 i) to every $\boldsymbol{x}$ vector, similar to what was done in Equation 4-3. Let $\boldsymbol{w}_k$ be the vector of weights (with the bias appended) for computing the value fed to output neuron k. The real part of the activation is often called the *logit* and is defined in Equation 4-34. In our context, $\boldsymbol{x}$ is the vector of activations of the final hidden layer, with (1.0 + 0 i) appended for the bias term.

$$\boldsymbol{logit}_k = \mathbf{Re}\left[\boldsymbol{w}_k \boldsymbol{\cdot} \boldsymbol{x}\right] \tag{4-34}$$

Suppose there are K classes. The model's estimated probability that the case that produced $\boldsymbol{x}$ belongs to class k (the activation of output neuron k) is given by Equation 4-35. These are called *SoftMax* activations because they are a smoother version of "hard max" winner-takes-all classification in which the model simply chooses the class that has maximum activation. These output activations are non-negative and sum to one.

$$p(y=k)=\frac{e^{logit_k}}{\sum_{i=1}^{K} e^{logit_k}} \tag{4-35}$$

Also, to conform to more general forms of the log likelihood function that you may encounter in more advanced texts, as well as to conform to the expression of the derivative that will soon be discussed, we express the log likelihood of a case in a seemingly complicated manner. For a given training case, define t_k as 1.0 if this case is a member of class k, and 0.0 otherwise. Also, define p_k as the activation of output neuron k, as given by Equation 4-35. Then the log of the likelihood corresponding to the model's parameters is given by Equation 4-36. This equation is called the *cross entropy*, and you can look up this term for some fascinating insights.

$$L=\sum_{k=1}^{K} t_k \log(p_k) \tag{4-36}$$

Observe that in the inner summation over classes, every term is zero except the term corresponding to the correct class. Thus, the log likelihood is just the log of the model's computed probability for the correct class of the case. Here are some observations about the log likelihood:

- Because p cannot exceed 1, the log likelihood is nonpositive.
- The better the model is at computing the correct class probabilities, the larger (closer to zero) this quantity will be, since it is the log probability of the correct class, and a good model will provide a large probability for the correct class.
- If the model is perfect, meaning that the computed probability of the correct class will be 1.0 for every case, the log likelihood will be zero, its maximum possible value.

Now comes a bit of almost unbelievable luck. Recall that Equation 4-32 led us to the derivative of the MSE with respect to the weighted sum coming into output neuron *k*. This was trivial calculus. Just imagine how horrendously complicated will be the formula for the derivative of the log likelihood shown in Equation 4-36, especially given the complexity of the probabilities defined in Equation 4-35 and its two equivalents. But here's the surprise. Without going through the considerable number of steps, we state that this derivative of Equation 4-36 for a case is given by Equation 4-37. Yup. It's that simple.

$$\delta_{or} = \frac{\partial L}{\partial logit_k} = p_k - t_k \tag{4-37}$$

Amazingly, except for a factor of two, the delta for a SoftMax output layer and maximum likelihood optimization is identical to that for a linear output layer and mean-squared-error optimization of the real part. All subsequent equations for the gradient hold. So, these two very different approaches to modeling can be handled with almost the same code.

Gradient of the Hidden-Layer Weights

With the preceding derivation tucked under our belts, it is time to examine the partial derivative of the error with respect to weights connecting the input layer to the hidden layer (or, by recursion discussed later, the weights connecting two hidden layers). It will now be necessary to peel more layers from the onion. It is the same onion, though, and the technique is similar to the one just described for the output layer. As before, we start by invoking the chain rule. Remember that the weight, *w*, now refers to a connection from the input layer to the hidden layer, as shown in Equation 4-38.

$$\begin{aligned} \frac{\partial E_{TOT}}{\partial w_r} &= \frac{\partial E_{TOT}}{\partial hnet_r}\frac{\partial hnet_r}{\partial w_r} + \frac{\partial E_{TOT}}{\partial hnet_i}\frac{\partial hnet_i}{\partial w_r} \\ \frac{\partial E_{TOT}}{\partial w_i} &= \frac{\partial E_{TOT}}{\partial hnet_r}\frac{\partial hnet_r}{\partial w_i} + \frac{\partial E_{TOT}}{\partial hnet_i}\frac{\partial hnet_i}{\partial w_i} \end{aligned} \tag{4-38}$$

Look back at Equations 4-30 and 4-31. We can find the right-hand factors in Equation 4-38 by analogy. They are shown in Equation 4-39.

$$\begin{aligned}\frac{\partial hnet_r}{\partial w_r} &= in_r \qquad \frac{\partial hnet_r}{\partial w_i} = -in_i \\ \frac{\partial hnet_i}{\partial w_r} &= in_i \qquad \frac{\partial hnet_i}{\partial w_i} = in_r\end{aligned} \tag{4-39}$$

Apply the chain rule to the left-hand factors. Once again we will label those terms to correspond to the δ (delta) terms in other standard real-domain derivations. This is shown in Equation 4-40.

$$\begin{aligned}\delta_{hr} &= \frac{\partial E_{TOT}}{\partial hnet_r} = \frac{\partial E_{TOT}}{\partial h_r}\frac{\partial h_r}{\partial hnet_r} + \frac{\partial E_{TOT}}{\partial h_i}\frac{\partial h_i}{\partial hnet_r} \\ \delta_{hi} &= \frac{\partial E_{TOT}}{\partial hnet_i} = \frac{\partial E_{TOT}}{\partial h_r}\frac{\partial h_r}{\partial hnet_i} + \frac{\partial E_{TOT}}{\partial h_i}\frac{\partial h_i}{\partial hnet_i}\end{aligned} \tag{4-40}$$

The right-hand factors in Equation 4-40 are the partial derivatives of the hidden-neuron activation function. Those derivatives were discussed on page 123.

The left-hand factors in that equation are considerably more difficult than they were for the output neurons. This is because the network error, E_{TOT}, is affected by the hidden neuron's activation, h, through all of the output neurons. This is illustrated in Figure 4-10.

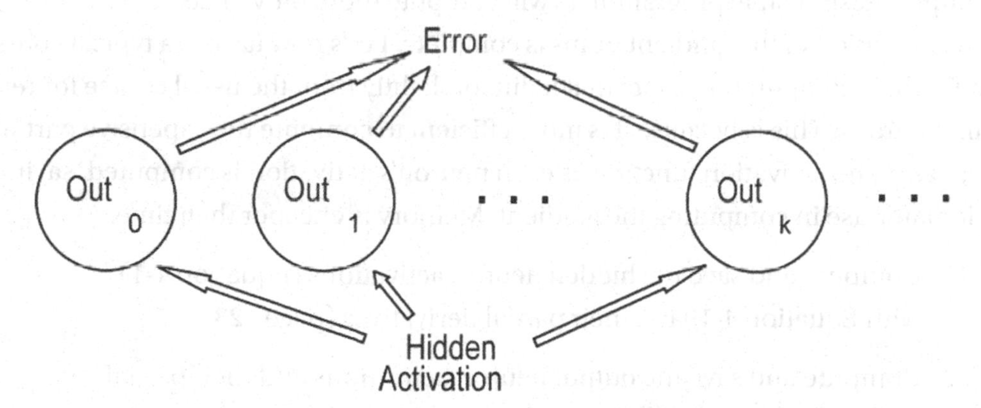

Figure 4-10. *Hidden neuron weights impact error though all outputs*

The chain rule as stated in Equation 4-27 is for just two functions. We now generalize it to as many functions as there are output neurons. Unfortunately, we must introduce a new subscript, k, for $anet_r$ and $anet_i$ to indicate which output neuron they represent.

Note that the summations in Equation 4-41 have nothing to do with the fact that our network error is the sum of the individual output neuron errors. They are a direct consequence of the chain rule. Even if more complicated network error functions are used, this equation still holds true.

$$\frac{\partial E_{TOT}}{\partial h_r} = \sum_k \left(\frac{\partial E_{TOT}}{\partial anet_{kr}} \frac{\partial anet_{kr}}{\partial h_r} \right) + \sum_k \left(\frac{\partial E_{TOT}}{\partial anet_{ki}} \frac{\partial anet_{ki}}{\partial h_r} \right)$$

$$\frac{\partial E_{TOT}}{\partial h_i} = \sum_k \left(\frac{\partial E_{TOT}}{\partial anet_{kr}} \frac{\partial anet_{kr}}{\partial h_i} \right) + \sum_k \left(\frac{\partial E_{TOT}}{\partial anet_{ki}} \frac{\partial anet_{ki}}{\partial h_i} \right) \tag{4-41}$$

Look back at Equation 4-30. We can now easily write the right-hand factors in each of the terms in Equation 4-41. This is done in Equation 4-42. Note that we must subscript the weights so that we know to which output neuron they lead.

$$\frac{\partial anet_{kr}}{\partial h_r} = w_{kr}, \quad \frac{\partial anet_{kr}}{\partial h_i} = -w_{ki}$$

$$\frac{\partial anet_{ki}}{\partial h_r} = w_{ki}, \quad \frac{\partial anet_{ki}}{\partial h_i} = w_{kr} \tag{4-42}$$

The left-hand factors in the summations in Equation 4-41 should look a little familiar. They are the output deltas already computed in Equation 4-32. There is one for each output. Responsible programmers will compute them only once.

The derivation of the gradient terms is complete. Let's now lay out a typical course of action for their computation. Note that it differs slightly from the usual course for real-domain neurons. This is because it is more efficient to compute the expensive partial derivatives of the activation function as each neuron's activation is computed, saving them for later use in computing the gradient. Memory is cheaper than time.

1. Compute and save the hidden neuron activations (Equation 4-1) with Equation 4-19and their partial derivatives (page 123).

2. Compute and save the output neuron activations and their partial derivatives in the same way as for the hidden neurons.

3. Compute the network error and its partial derivative with respect to output activations using Equation 4-33 or whatever is appropriate for the error measure (for example, Equation 4-37 for SoftMax outputs).

4. Compute delta for each output neuron using Equation 4-32. When the outputs are linear, as is the case for the DEEP program and throughout this text, this delta is equal to the quantities computed in the prior step because of the right-hand factors in that equation being identically 1.0 or 0.0 as discussed earlier.
5. Compute the output-layer gradient using Equation 4-29.
6. For each hidden neuron:
    a. Cumulate the product of all output deltas times the weight connecting that output to the hidden neuron being done (Equation 4-41).
    b. Compute this hidden neuron's delta using Equation 4-40. If there is another hidden layer, save delta.
    c. Compute this hidden neuron's gradient using Equation 4-38.

Repeat step 6 for any additional hidden layers, each time using the delta vector computed for the previously processed hidden layer.

Code for Gradient Computation

This section presents some subroutines and code fragments that illustrate gradient computation. The fragments especially should be seen as templates for developing your own application-specific code. Note that subroutine `activity_cc()` was discussed on page 124. Subroutine `activity()` is just the simple real-domain version, and it is included in the set of code that can be downloaded for free from my web site.

Evaluating the Entire Network and Derivatives

The first subroutine propagates data through an entire neural network, from raw inputs to output. It can handle both strictly real and fully complex models. The calling parameter list and variable declarations are as follows:

```
void trial_thr (
  double *input ,         // Input vector nin long
  int n_layers ,          // Number of layers, including output, not including input
  int nin ,               // Number of possibly complex inputs to the model
```

```
  double *outputs ,          // Output vector of the model
  int nout ,                 // Number of possibly complex outputs
  int *nhid ,                // nhid[i] is the number of hidden neurons in hidden layer i
  double *weights[] ,        // weights[i] points to the weight vector for hidden layer i
  double *hid_act[] ,        // hid_act[i] points to the vector of activations of hidden layer i
  double *hid_rr[] ,         // Partial of real activation wrt real input
  double *hid_ii[] ,         // Ditto, imaginary
  double *hid_ri[] ,         // Ditto, real wrt imaginary = imaginary wrt real
  double *last_layer_weights , // Weights of final layer
  int complex ,              // Is this a complex network?
  int classifier             // If nonzero use SoftMax output; else use linear output
  )
{
  int i, ilayer ;
  double sum ;
```

We process layers one at a time. The first possibility is that there are no hidden layers; the input layer connects directly to the output layer, making this an ordinary linear regression model, possibly complex.

```
  for (ilayer=0 ; ilayer<n_layers ; ilayer++) {  // For all layers, input to output

    if (ilayer == 0 && n_layers == 1) {           // Direct input to output?
      for (i=0 ; i<nout ; i++) {                  // For all output neurons

        if (complex)
          activity_cc ( input , last_layer_weights+i*2*(nin+1) , outputs+2*i ,
                        NULL , NULL , NULL , nin , 1 ) ;
        else
          activity ( input , last_layer_weights+i*(nin+1) , outputs+i , nin , 1 ) ;
        }
      }
```

Another possibility is that this is the first hidden layer, so the inputs to this layer are the raw inputs to the model.

```
   else if (ilayer == 0) {              // First hidden layer?
    for (i=0 ; i<nhid[ilayer] ; i++) { // For all neurons in this hidden layer

      if (complex) {
        if (hid_rr != NULL)            // Does the caller want derivatives?
          activity_cc ( input , weights[ilayer]+i*2*(nin+1) , hid_act[ilayer]+2*i ,
                      hid_rr[ilayer]+i , hid_ii[ilayer]+i , hid_ri[ilayer]+i , nin , 0 ) ;
        else
          activity_cc ( input , weights[ilayer]+i*2*(nin+1) , hid_act[ilayer]+2*i ,
                      NULL , NULL , NULL , nin , 0 ) ;
        }

      else                             // It's a real-domain model
        activity ( input , weights[ilayer]+i*(nin+1) , hid_act[ilayer]+i , nin , 0 ) ;
      }
    }
```

Another possibility is that this is a hidden layer that follows a hidden layer. In this case, the inputs to this layer are the activations of the preceding hidden layer.

```
   else if (ilayer < n_layers-1) {          // Subsequent hidden layer?
    for (i=0 ; i<nhid[ilayer] ; i++) {      // For all neurons in this layer

      if (complex) {
        if (hid_rr != NULL)                 // Does the caller want derivatives?
          activity_cc ( hid_act[ilayer-1] , weights[ilayer]+i*2*(nhid[ilayer-1]+1) ,
                      hid_act[ilayer]+2*i , hid_rr[ilayer]+i ,
                      hid_ii[ilayer]+i , hid_ri[ilayer]+i , nhid[ilayer-1] , 0 );
        else
          activity_cc ( hid_act[ilayer-1] , weights[ilayer]+i*2*(nhid[ilayer-1]+1) ,
                      hid_act[ilayer]+2*i , NULL , NULL , NULL , nhid[ilayer-1] , 0 );
        }

      else                                  // Real-domain model
        activity ( hid_act[ilayer-1] , weights[ilayer]+i*(nhid[ilayer-1]+1) ,
                 hid_act[ilayer]+i , nhid[ilayer-1] , 0 );
      }
    }
```

Finally, we may have reached the output layer.

```
else {                                              // Final layer
  for (i=0 ; i<nout ; i++) {                        // For all output neurons

    if (complex)
      activity_cc ( hid_act[ilayer-1] , last_layer_weights+i*2*(nhid[ilayer-1]+1) ,
                    outputs+2*i , NULL , NULL , NULL , nhid[ilayer-1] , 1 );
    else
      activity ( hid_act[ilayer-1] , last_layer_weights+i*(nhid[ilayer-1]+1) ,
                 outputs+i , nhid[ilayer-1] , 1 );
    } // For i
  } // Else output layer
} // For all layers
```

If this is a classifier, we need to do one more thing: the real parts of the outputs must be converted to SoftMax per Equation 4-35. The imaginary parts for a complex-domain model are completely ignored for SoftMax conversion of the outputs.

```
if (classifier) { // Classifier is always SoftMax

  if (complex) {
    sum = 0.0 ;
    for (i=0 ; i<nout ; i++) { // For all outputs
      if (outputs[2*i] < 300.0)             // Cheap insurance against problems
        outputs[2*i] = exp ( outputs[2*i] ) ;
      else
        outputs[2*i] = exp ( 300.0 ) ;
      sum += outputs[2*i] ;
      }
    for (i=0 ; i<nout ; i++)
      outputs[2*i] /= sum ;                 // Ignore imaginary parts
    }

  else {      // Real domain
    sum = 0.0 ;
```

```
      for (i=0 ; i<nout ; i++) { // For all outputs
         if (outputs[i] < 300.0)
            outputs[i] = exp ( outputs[i] ) ;
         else
            outputs[i] = exp ( 300.0 ) ;
         sum += outputs[i] ;
         }
      for (i=0 ; i<nout ; i++)
         outputs[i] /= sum ;
      }
   } // If classifier
}
```

Computing the Gradient

Numerical optimization routines will need the gradient often, so we must be able to compute it quickly and accurately. Moreover, they will want the gradient to be stored as a single vector. This is facilitated by our use of an array of pointers: grad_ptr[ilayer] points to the location in this grand gradient vector of the start of the gradient for layer ilayer (with 0 being the first hidden layer), which simplifies addressing in the routine that follows. Here is the calling parameter list and variable declarations:

```
double batch_gradient (
   int istart ,             // Index of starting case in input matrix
   int istop ,              // And one past last case
   double *input ,          // Input matrix; each case is max_neurons long
   double *targets ,        // Target matrix; strictly real, so each case is nout long
   int *class_ids ,         // Class id vector if classifier (ignored if not)
   int n_layers ,           // Number of layers, including output, not including input
   int n_weights ,          // Total n of weights, including final layer and all bias terms
   int nin ,                // Number of possibly complex inputs to the model
   double *outputs ,        // Output vector of the model; used as work vector here
   int nout ,               // Number of possibly complex outputs (targets)
   int *nhid ,              // nhid[i] is the number of hidden neurons in hidden layer i
   double *weights[] ,      // weights[i] points to the weight vector for hidden layer i
   double *hid_act[] ,      // hid_act[i] points to the vector of activations of hidden layer i
```

```
  double *hid_rr[] ,           // Partial of real activation wrt real input
  double *hid_ii[] ,           // Ditto, imaginary
  double *hid_ri[] ,           // Ditto, real wrt imaginary = imaginary wrt real
  int max_neurons ,            // Number of columns in input matrix
  double *this_delta ,         // Delta for the current layer
  double *prior_delta ,        // And saved for use in the prior (next to be processed) layer
  double **grad_ptr ,          // grad_ptr[i] points to gradient for layer i
  double *last_layer_weights , // Weights of final layer
  double *grad ,               // All computed gradients, strung out as a single long vector
  int complex ,                // Is this a complex network?
  int classifier               // If nonzero use SoftMax output; else use linear output
  )
{
  int i, j, icase, ilayer, nprev, nthis, nnext, mult, iclass ;
  double diff, *dptr, error, *targ_ptr, *prevact, *gradptr ;
  double rsum, isum, delta, rdelta, idelta, *nextcoefs, tval ;
  double *rr_ptr, *ii_ptr, *ri_ptr ;
```

The variable mult will be 2 for complex models and 1 for real models. This will prove convenient often. We'll sum the gradient for a block of cases, so begin by zeroing the summation area and the variable where we will cumulate the error to be minimized. Within the case loop, find the current case and call trial_thr() to pass through the entire network, computing all activations and derivatives of the activation function.

```
  mult = complex ? 2 : 1 ;             // Numbers per neuron

  for (i=0 ; i<n_weights ; i++)        // Zero gradient for summing
    grad[i] = 0.0 ;                    // All layers are strung together here

  error = 0.0 ;                        // Will cumulate total error here

  for (icase=istart ; icase<istop ; icase++) {

    dptr = input + icase * max_neurons ; // Point to this sample; max_neurons is col dim
    trial_thr ( dptr , n_layers , nin , outputs , nout , nhid ,
              weights , hid_act , hid_rr , hid_ii , hid_ri ,
              last_layer_weights , complex , classifier ) ;
```

There are now three possibilities for the nature of the output layer.

1) This is a classification model, in which case outputs are SoftMax.

2) We are training the final model. Targets are strictly real, so if this is a complex model, we ignore the imaginary part of the prediction. The user has specified a target vector.

3) We are training an autoencoder. The targets will be the inputs, and if this is a complex model, we must take both the real and imaginary parts into account for error computation.

These three possibilities are handled as shown on the next page. The deltas are computed using Equation 4-32. But recall that, as discussed on that page, the outputs are linear. As a result, many of the terms in that fierce-looking equation are always 1.0 or 0.0, leading to the vast simplification seen in this code. Finally, we are computing the *negative* derivative, so many signs will be flipped relative to the equations.

```
if (classifier) {                                  // SoftMax
  iclass = class_ids[icase] ;                      // This is the true class

  for (i=0 ; i<nout ; i++) {                       // Compute delta for each output
    tval = (i == iclass) ? 1.0 : 0.0 ;             // Equation 4-37
    this_delta[mult*i] = tval - outputs[mult*i] ; // Neg deriv of cross entropy wrt input
    if (complex)
      this_delta[2*i+1] = 0.0 ;                    // The imaginary part is ignored
    }
  error -= log ( outputs[mult*iclass] + 1.e-30 ) ; // Equation 4-36
  }

else if (targets != NULL) {                        // Training final model
  targ_ptr = targets + icase * nout ;              // Targets are strictly real

  for (i=0 ; i<nout ; i++) {                       // For each output
    diff = outputs[mult*i] - targ_ptr[i] ;         // Real part of prediction compared to target
    error += diff * diff ;                         // Squared error
    this_delta[mult*i] = -2.0 * diff ;             // Neg deriv of squared error wrt input
    if (complex)                                   // Above is real part of Equation 4-32
      this_delta[2*i+1] = 0.0 ;                    // Target is real so ignore imaginary part
    }
  }
```

```
else {                                           // Training an autoencoder
  targ_ptr = input + icase * max_neurons ; // Point to this case; input is target

  for (i=0 ; i<mult*nout ; i++) {                // For each output, real and imaginary parts
    diff = outputs[i] - targ_ptr[i] ;
    error += diff * diff ;
    this_delta[i] = -2.0 * diff ;                  // Equation 4-32
    }
  }
```

Now that we know the output layer delta vectors, we can compute the gradient for the output weights.

```
if (n_layers == 1) {               // No hidden layer
  nprev = nin ;                    // Number of (possibly complex) inputs
  prevact = input + icase * max_neurons ; // Point to this sample
  }
else {
  nprev = nhid[n_layers-2] ;             // n_layers-2 is the last hidden layer
  prevact = hid_act[n_layers-2] ;        // Point to layer feeding the output layer
  }

gradptr = grad_ptr[n_layers-1] ;         // Output gradient in grand grad vector
for (i=0 ; i<nout ; i++) {               // For all output neurons

  if (complex) {
    rdelta = this_delta[2*i] ;
    idelta = this_delta[2*i+1] ;
    for (j=0 ; j<nprev ; j++) {          // Eq (4-31) on Page 129 for Eq (4-29)
      *gradptr++ += rdelta * prevact[2*j] + idelta * prevact[2*j+1] ;
      *gradptr++ += -rdelta * prevact[2*j+1] + idelta * prevact[2*j] ;
      }
    *gradptr++ += rdelta ;               // Bias activation is always 1
    *gradptr++ += idelta ;
    }
```

```
      else {
        delta = this_delta[i] ;                     // Neg deriv of criterion wrt logit
        for (j=0 ; j<nprev ; j++)
           *gradptr++ += delta * prevact[j] ; // Cumulate for all training cases
        *gradptr++ += delta ;                     // Bias activation is always 1
        }
      }

   nnext = nout ;                               // Prepare for moving back one layer
   nextcoefs = last_layer_weights ;
```

The easy part is done. Now we must move back to the hidden layer(s). In this step we have to deal with the awkward fact that the weight for any hidden neuron impacts the error through every output neuron, as depicted in Figure 4-10. Thus, we will need to sum these impacts.

```
   for (ilayer=n_layers-2 ; ilayer>=0 ; ilayer--) {  // For each hidden layer, backwards
      nthis = nhid[ilayer] ;                 // Number of neurons in this hidden layer
      gradptr = grad_ptr[ilayer] ;           // Point to gradient for this layer

      if (complex) {                         // Do we need complex derivatives?
         rr_ptr = hid_rr[ilayer] ;           // For real models, we compute derivatives now
         ii_ptr = hid_ii[ilayer] ;           // But for complex models we precomputed them
         ri_ptr = hid_ri[ilayer] ;           // with activations and saved them
         }

      for (i=0 ; i<nthis ; i++) {            // For each neuron in this layer

         if (complex) {

            rsum = isum = 0.0 ;              // For Equation 4-41
            for (j=0 ; j<nnext ; j++) {      // Summation depicted in Figure 4-10
               rsum += this_delta[2*j]   * nextcoefs[j*2*(nthis+1)+2*i] +
                       this_delta[2*j+1] * nextcoefs[j*2*(nthis+1)+2*i+1] ;
               isum += -this_delta[2*j]  * nextcoefs[j*2*(nthis+1)+2*i+1] +
                       this_delta[2*j+1] * nextcoefs[j*2*(nthis+1)+2*i] ;
               }
```

```
        rdelta = rsum * rr_ptr[i] + isum * ri_ptr[i] ;        // Equation 4-40
        idelta = rsum * ri_ptr[i] + isum * ii_ptr[i] ;
        prior_delta[2*i] = rdelta ;                           // Save it for the next layer back
        prior_delta[2*i+1] = idelta ;
```

We are now ready to use Equation 4-38 to compute the gradient entry for this hidden neuron. The if (ilayer==0)...else blocks are identical except that if it's the first hidden layer, then it is being fed by the model's raw inputs, while if it's a subsequent hidden layer, then it is being fed by the prior hidden layer. The computation is identical in both cases.

```
    if (ilayer == 0) {                        // First hidden layer?
      prevact = input + icase * max_neurons ; // Point to this sample
      for (j=0 ; j<nin ; j++) {
        *gradptr++ += rdelta * prevact[2*j] + idelta * prevact[2*j+1] ;
        *gradptr++ += -rdelta * prevact[2*j+1] + idelta * prevact[2*j] ;
        }
      }

    else {    // There is at least one more hidden layer prior to this one
      prevact = hid_act[ilayer-1] ;
      for (j=0 ; j<nhid[ilayer-1] ; j++) {
        *gradptr++ += rdelta * prevact[2*j] + idelta * prevact[2*j+1] ;
        *gradptr++ += -rdelta * prevact[2*j+1] + idelta * prevact[2*j] ;
        }
      }

    *gradptr++ += rdelta ;    // Bias activation is always 1+0i
    *gradptr++ += idelta ;
    } // Complex model
```

The previous code handled complex-domain models. The following code is identical except that it handles the much simpler real-domain models. Also note that for complex models we computed and saved the partial derivatives of the hidden neuron activation functions at the same time that we computed their activations. But the derivatives of the real-domain activations are so simple (Equation 4-25) that we just compute them here. This code will not be commented because every operation is an exact analog of the complex-domain computations just presented.

```
      else { // Real
        delta = 0.0 ;
        for (j=0 ; j<nnext ; j++)
          delta += this_delta[j] * nextcoefs[j*(nthis+1)+i] ;
        delta *= hid_act[ilayer][i] * (1.0 - hid_act[ilayer][i]) ; // Derivative
        prior_delta[i] = delta ;                    // Sav e it for the next layer back
        if (ilayer == 0) {                          // First hidden layer?
          prevact = input + icase * max_neurons ; // Point to this sample
          for (j=0 ; j<nin ; j++)
            *gradptr++ += delta * prevact[j] ;
          }

        else {      // There is at least one more hidden layer prior to this one
          prevact = hid_act[ilayer-1] ;
          for (j=0 ; j<nhid[ilayer-1] ; j++)
            *gradptr++ += delta * prevact[j] ;
          }
        *gradptr++ += delta ;   // Bias activation is always 1
        }

      } // For all neurons in this hidden layer
```

We have computed the gradient terms for every neuron in this hidden layer. We also saved in prior_delta thc deltas (Equation 4-40). Copy these deltas for use in the next hidden layer back. Also set nnext and nextcoefs to prepare for the nasty summations that are required for hidden layers.

```
    for (i=0 ; i<mult*nthis ; i++)       // These will be delta for the next layer back
      this_delta[i] = prior_delta[i] ;

    nnext = nhid[ilayer] ;               // Prepare for the next layer back
    nextcoefs = weights[ilayer] ;
    } // For all layers, working backwards

  } // for all cases

  return error ; // MSE or negative log likelihood
}
```

Multithreading Gradient Computation

Volume I of this series contained extensive details on the mechanics of my favorite method for multithreading computations in Windows. For this reason, detailed treatment of background material and motivation will be omitted here. Rather, this section will list the core elements of source code for computing the gradient, along with explanations that seem particularly important.

We must have a data structure to pass parameters and pointers to the threaded routine. Here it is:

```
typedef struct {
  int istart ;
  int istop ;
  int complex ;
  int classifier ;
  int n_layers ;
  int n_weights ;
  int nin ;
  int nout ;
  int *nhid ;
  int max_neurons ;
  double *input ;
  double *targets ;
  int *class_ids ;
  double *outputs ;
  double **weights ;
  double **hid_act ;
  double **hid_rr ;
  double **hid_ii ;
  double **hid_ri ;
  double *this_delta ;
  double *prior_delta ;
  double **grad_ptr ;
  double *last_layer_weights ;
```

```
   double *grad ;
   double error ;
   } GRAD_THR_PARAMS ;
```

The wrapper routine, which is launched as a thread and calls the core computation routine, is as follows:

```
static unsigned int __stdcall batch_gradient_wrapper ( LPVOID dp )
{
((GRAD_THR_PARAMS *) dp)->error = batch_gradient (
                  ((GRAD_THR_PARAMS *) dp)->istart ,
                  ((GRAD_THR_PARAMS *) dp)->istop ,
                  ((GRAD_THR_PARAMS *) dp)->input ,
                  ((GRAD_THR_PARAMS *) dp)->targets ,
                  ((GRAD_THR_PARAMS *) dp)->class_ids ,
                  ((GRAD_THR_PARAMS *) dp)->n_layers ,
                  ((GRAD_THR_PARAMS *) dp)->n_weights ,
                  ((GRAD_THR_PARAMS *) dp)->nin ,
                  ((GRAD_THR_PARAMS *) dp)->outputs ,
                  ((GRAD_THR_PARAMS *) dp)->nout ,
                  ((GRAD_THR_PARAMS *) dp)->nhid ,
                  ((GRAD_THR_PARAMS *) dp)->weights ,
                  ((GRAD_THR_PARAMS *) dp)->hid_act ,
                  ((GRAD_THR_PARAMS *) dp)->hid_rr ,
                  ((GRAD_THR_PARAMS *) dp)->hid_ii ,
                  ((GRAD_THR_PARAMS *) dp)->hid_ri ,
                  ((GRAD_THR_PARAMS *) dp)->max_neurons ,
                  ((GRAD_THR_PARAMS *) dp)->this_delta ,
                  ((GRAD_THR_PARAMS *) dp)->prior_delta ,
                  ((GRAD_THR_PARAMS *) dp)->grad_ptr ,
                  ((GRAD_THR_PARAMS *) dp)->last_layer_weights ,
                  ((GRAD_THR_PARAMS *) dp)->grad ,
                  ((GRAD_THR_PARAMS *) dp)->complex ,
                  ((GRAD_THR_PARAMS *) dp)->classifier ) ;
   return 0 ;
}
```

The routine that is called to oversee gradient computation breaks the training set into separate batches, each of which is processed by a separate thread. Its calling parameter list and variable declarations are shown on the next page. Note that unlike the prior routines, this one is not self-contained; it references numerous CpxAuto class variables that are unique to my implementation in the DEEP program. Therefore, you will need to adapt this code to your own program architecture. The idea here is to provide a template for multithreaded computation.

```
double CpxAuto::gradient_thr (
   int nc ,                   // Number of cases
   int nin ,                  // Number of possibly complex inputs
   double *input ,            // Nc by max_neurons input matrix
   int nout ,                 // Number of possibly complex outputs
   double *target ,           // Nc by nout target matrix, or autoencoding if NULL
   int n_layers ,             // Number of layers
   int *nhid ,                // Number of hidden neurons in each layer
   int n_weights ,            // Total number of weights, including final layers and bias
   double *weights[] ,        // Weight matrices for layers
   int use_final_layer_weights , // Use final_layer_weights (vs last weight layer)?
   double *grad               // Concatenated gradient vector, which is computed here
   )
{
   int i, j, ilayer, ineuron, ivar, n, istart, istop, n_done, ithread, mult ;
   int n_in_batch, n_threads, ret_val, nin_this_layer, n_last_layer_weights ;
   double error, *wptr, *gptr, factor, *hid_act_ptr[MAX_THREADS][MAX_LAYERS] ;
   double *grad_ptr_ptr[MAX_THREADS][MAX_LAYERS] ;
   double *hid_rr_ptr[MAX_THREADS][MAX_LAYERS] ;
   double *hid_ii_ptr[MAX_THREADS][MAX_LAYERS] ;
   double *hid_ri_ptr[MAX_THREADS][MAX_LAYERS] ;
   double wpen, *last_layer_weights ;
   char msg[256] ;
   GRAD_THR_PARAMS params[MAX_THREADS] ;
   HANDLE threads[MAX_THREADS] ;
```

As in the core gradient routine, we set mult to be 2 for a complex model and 1 for a real model. This will prove handy. We scale the weight penalty to keep it independent of the model size.

```
mult = is_complex ? 2 : 1 ;
wpen = TrainParams.wpen / n_weights ;
```

We saw earlier that the gradients for all layers are strung together into a single grand gradient vector, with the starting point for each layer being kept in grad_ptr[]. On the next page we see how this array of pointers is built.

```
gptr = grad ; // Allocated n_weights * max_threads long

for (ilayer=0 ; ilayer<n_layers ; ilayer++) {
  grad_ptr[ilayer] = gptr ;

  if (ilayer == 0 && n_layers == 1) {          // Direct input to output?
    n = nout * mult * (nin+1) ;                // This many inputs to each neuron in layer
    gptr += n ;                                // Not needed, but it illustrates the process
    }

  else if (ilayer == 0) {                      // First hidden layer?
    n = nhid[ilayer] * mult * (nin+1) ;        // This many inputs to each neuron in layer
    gptr += n ;
    }

  else if (ilayer < n_layers-1) {              // Subsequent hidden layer?
    n = nhid[ilayer] * mult * (nhid[ilayer-1]+1) ; // This many ins to each neuron in layer
    gptr += n ;
    }

  else                                         // Output layer
    n = nout * mult * (nhid[ilayer-1]+1) ;     // This many inputs to each neuron in layer
                                               // Line not needed but here for clarity
  } // For all layers, including output
```

Many of the parameters passed to the core gradient routine can be preset easily because they are constant for every batch. Do so now.

```
for (i=0 ; i<max_threads ; i++) {
   params[i].input = input ;
   params[i].targets = target ;           // Will be NULL for autoencoding
   params[i].class_ids = class_ids ;
   params[i].n_layers = n_layers ;
   params[i].n_weights = n_weights ;
   params[i].nin = nin ;
   params[i].nout = nout ;
   params[i].nhid = nhid ;
   params[i].max_neurons = max_neurons ;
   params[i].weights = weights ;
   params[i].last_layer_weights = last_layer_weights ;
   // Outputs is used strictly for scratch in each thread, not for saving predictions
   if (use_final_layer_weights)           // Are we training the final model?
      params[i].outputs = outputs + i * mult * nout ;
   else
      params[i].outputs = autoencode_out + i * mult * nin ; // Autoencoding layer

   // Each thread must have its own private copy of these work areas
   params[i].this_delta = this_layer + i * max_neurons ;
   params[i].prior_delta = prior_layer + i * max_neurons ;
   params[i].grad = grad + i * n_weights ;

   for (j=0 ; j<n_layers ; j++) {
      hid_act_ptr[i][j] = hid_act[j] + i * max_neurons ;
      grad_ptr_ptr[i][j] = grad_ptr[j] + i * n_weights ;
      if (is_complex) {
         hid_rr_ptr[i][j] = hid_rr[j] + i * max_neurons / 2 ;          // These are real
         hid_ii_ptr[i][j] = hid_ii[j] + i * max_neurons / 2 ;          // So must divide by 2
         hid_ri_ptr[i][j] = hid_ri[j] + i * max_neurons / 2 ;          // to get actual count
         }
      }

   params[i].hid_act = hid_act_ptr[i] ;
   params[i].grad_ptr = grad_ptr_ptr[i] ;
```

```
   if (is_complex) {
      params[i].hid_rr = hid_rr_ptr[i] ;
      params[i].hid_ii = hid_ii_ptr[i] ;
      params[i].hid_ri = hid_ri_ptr[i] ;
      }
   else
      params[i].hid_rr = params[i].hid_ii = params[i].hid_ri = NULL ;

   params[i].complex = is_complex ;

   if (target == NULL)              // Autoencoding is never a classifier (of course!)
      params[i].classifier = 0 ;
   else
      params[i].classifier = classifier ;
   }
```

The training set is divided into equal-size batches, with each given to a separate thread. The algorithm for doing so is simple: for each available thread, divide the number of training cases left to do by the number of threads that remain available and launch a batch of that size.

```
n_threads = max_threads ;            // Try to use as many as possible
if (nc / n_threads < 100)            // But because threads have overhead
   n_threads = 1 ;                   // Avoid using them if the batch is small

istart = 0 ;       // Batch start = training data start
n_done = 0 ;       // Number of training cases done in this epoch so far

for (ithread=0 ; ithread<n_threads ; ithread++) {
   n_in_batch = (nc - n_done) / (n_threads - ithread) ; // Cases left to do / batches left
   istop = istart + n_in_batch ;                         // Stop just before this index

   // Set the pointers that vary with the batch

   params[ithread].istart = istart ;
   params[ithread].istop = istop ;

   // Launch a thread for this batch
   threads[ithread] = (HANDLE) _beginthreadex ( NULL , 0 , batch_gradient_wrapper ,
                                     &param s[ithread] , 0 , NULL ) ;
```

```
    if (threads[ithread] == NULL) { // Pathological... should never happen
      for (i=0 ; i<n_threads ; i++) {
        if (threads[i] != NULL)
          CloseHandle ( threads[i] ) ;
        }
      return -1.e40 ;                    // Flag the caller that disaster struck
      }

  n_done += n_in_batch ;                 // Tally the number of cases submitted so far
  istart = istop ;                       // Next batch starts here
  } // For all threads / batches
```

There's nothing to do now but wait for all those threads to finish. The timeout of 1200000 is arbitrary, but it must be large enough that no foreseeable job would exceed it. It also must not be so gigantic that a hapless user might wait forever for an impractically large problem.

```
  ret_val = WaitForMultipleObjects ( n_threads , threads , TRUE , 1200000 ) ;
  if (ret_val == WAIT_TIMEOUT || ret_val == WAIT_FAILED ||
     ret_val < 0 || ret_val >= n_threads) // Pathological; should never happen
    return -1.e40 ; // Flag user that disaster struck
```

The following loop cumulates the error and gradient from all the threads, adding them all into the results for thread zero.

```
  CloseHandle ( threads[0] ) ;
  for (ithread=1 ; ithread<n_threads ; ithread++) {
    params[0].error += params[ithread].error ;
    for (i=0 ; i<n_weights ; i++)
      params[0].grad[i] += params[ithread].grad[i] ;
    CloseHandle ( threads[ithread] ) ;
    }
```

At this time we have the pooled results for the complete training set in the error and gradient areas for thread zero. Find the mean per presentation and term. This scaling is not required; it's just convenient for the user. But whatever scaling is chosen, make sure it is the same for the error and the gradient! Also, note that grad and params[0].grad are the same! (Look back at the parameter initialization if this is not clear.) The gradient scaling is written this way just for clarity.

```
   factor = 1.0 / (nc * mult * nout) ;
   error = factor * params[0].error ;

   for (i=0 ; i<n_weights ; i++)
      grad[i] = factor * params[0].grad[i] ;
```

The final step is to apply the weight penalty. This was discussed in detail in Volume I of this series, so we won't dwell on it here. We use a squared-weight penalty, so its derivative is two times the weight. The first block shown here does the hidden layers, and the second block does the output layer.

```
   penalty = 0.0 ;

   nin_this_layer = nin ;
   for (ilayer=0 ; ilayer<n_layers-1 ; ilayer++) { // Do all hidden layers

      for (ineuron=0 ; ineuron<nhid[ilayer] ; ineuron++) {
         wptr = weights[ilayer] + ineuron * mult * (nin_this_layer+1) ;  // Weights
         gptr = grad_ptr[ilayer] + ineuron * mult * (nin_this_layer+1) ; // Gradient
         for (ivar=0 ; ivar<mult*nin_this_layer ; ivar++) {             // Do not include bias
            penalty += wptr[ivar] * wptr[ivar] ;
            gptr[ivar] -= 2.0 * wpen * wptr[ivar] ;
            }
         }
      nin_this_layer = nhid[ilayer] ;
      }

   for (ineuron=0 ; ineuron<nout ; ineuron++) {
      wptr = last_layer_weights + ineuron * n_last_layer_weights ;
      gptr = grad_ptr[n_layers-1] + ineuron * n_last_layer_weights ;
      for (ivar=0 ; ivar<mult*nin_this_layer ; ivar++) {             // Do not include bias
         penalty += wptr[ivar] * wptr[ivar] ;
         gptr[ivar] -= 2.0 * wpen * wptr[ivar] ;
         }
      }

   penalty *= wpen ;
   return error + penalty ;
}
```

CUDA Gradient Computation

Volume I of this series presented an introduction to CUDA computation, especially focusing on issues related to computing activations and gradients. For the sake of our vanishing forests, I will not repeat this material. Here, I assume that you have, at a minimum, read and digested that information. The focus now will be exclusively on extending those algorithms to the complex domain.

This section contains numerous code fragments that are used for explication of the algorithms. More complete versions of this code, with fragments shown in context and with more detail, can be downloaded for free from my web site. However, even that code has all error checking removed for clarity. Also, that code is intended to be used as templates and illustrations of the relevant equations rather than as directly usable "drop-in" code.

The Overall Algorithm

We begin with the routine that runs on the host machine and coordinates computation of the gradient. Its calling parameter list and variable declarations are as follows:

```
double CpxAuto::gradient_cuda (
   int nc ,                    // Number of cases
   int nin ,                   // Number of (possibly complex) inputs
   double *input ,             // Nc by max_neurons input matrix
   int nout ,                  // Number of (possibly complex) outputs
   double *target ,            // Nc by nout target matrix, or NULL if autoencoding
   int n_layers ,              // Number of layers, including output layer
   int *nhid ,                 // Number of hidden neurons in each layer
   int n_weights ,             // Total number of weights, including final layer and bias
   double *weights[] ,         // Weight matrices for layers
   int use_final_layer_weights , // Use final_layer_weights (vs last weight layer)?
   double *grad                // Concatenated gradient vector, which is computed here
   )
{
   int i, k, n, ilayer, ineuron, ivar, ret_val, ibatch, n_in_batch, n_subsets, istart, istop ;
   int n_done, max_batch, int n_prior, gradlen, nin_this_layer, timer ;
   int n_last_layer_weights, mult ;
   double mse, wpen, *wptr, *gptr, *last_layer_weights ;
```

This routine will be called in either of two situations:

- The model is being trained in supervised mode. It will be called with use_final_layer_weights nonzero (true), which tells the routine that the weights for the output layer are in the class's member array final_layer_weights. The caller must supply the supervisory targets in the target array, which is always real, even for complex-domain models. (The imaginary part of predictions is always ignored.)
- The model is an autoencoder, perhaps a single layer being trained greedily, or perhaps a set of layers being fine-tuned as a multilayer autoencoder. In this situation, use_final_layer_weights is zero (false), indicating that the output layer's weights are the last set of weights in the weights matrix. The target vector would be input NULL because the targets are the inputs. For a complex-domain model, the imaginary parts of the predictions are taken into consideration along with the real parts.

These two possibilities are handled in the following code. We also set mult to be 1 or 2 depending on whether this is a real or complex model.

```
mult = is_complex ? 2 : 1 ;

if (use_final_layer_weights) {                       // Supervised training
   assert ( target != NULL ) ;
   last_layer_weights = final_layer_weights ;
   n_last_layer_weights = n_final_layer_weights ;    // Per output, not total
   }

else {                                               // Greedily training autoencoder
   assert ( target == NULL ) ;                       // which may be complex
   last_layer_weights = weights[n_layers-1] ;
   n_last_layer_weights = mult * (nhid[n_layers-2] + 1) ;
   }
```

We set up the pointers to the gradient for each layer, exactly as we did for the threaded version.

```
gptr = grad ;

for (ilayer=0 ; ilayer<n_layers ; ilayer++) {
  grad_ptr[ilayer] = gptr ;

  if (ilayer == 0 && n_layers == 1) {          // Direct input to output?
    n = nout * mult * (nin+1) ;                // This many inputs to each neuron in layer
    gptr += n ;                                // Not needed, but it illustrates the process
    }

  else if (ilayer == 0) {                      // First hidden layer?
    n = nhid[ilayer] * mult * (nin+1) ;        // This many inputs to each neuron in layer
    gptr += n ;
    }

  else if (ilayer < n_layers-1) {              // Subsequent hidden layer?
    n = nhid[ilayer] * mult * (nhid[ilayer-1]+1) ; // This many inputs to each neuron
    gptr += n ;
    }

  else
    n = nout * mult * (nhid[ilayer-1]+1) ;     // Not needed but illustrates process
  } // For all layers, including output
```

The next step is a little tricky but critical. The length of the gradient vector can be enormous, and as will be seen later the CUDA routines compute and store it separately for each case in a batch (subset of the complete training set). If the batches are large, the number of bytes to be allocated on the CUDA device can easily exceed the maximum positive number representable in 32 bits, which can cause problems. Moreover, currently available CUDA devices have a hardware limit of 65535 on the size of the grid. These two limitations require that we impose an upper limit on the size of a batch, which is accomplished by imposing a lower limit on the number of subsets into which the training set is divided. Moreover, we must allow the user to further increase the number of subsets to avoid the infamous Windows display timeouts. This is done as follows:

```
gradlen = 0 ;                              // Counts length of gradient
n_prior = nin ;                            // Size of feed into a layer

for (i=0 ; i<n_layers-1 ; i++) {           // Hidden layers
  gradlen += mult * nhid[i] * (n_prior + 1) ; // +1 to include bias
  n_prior = nhid[i] ;
  }

gradlen += mult * nout * (n_prior + 1) ;   // Output layer
assert ( gradlen == n_weights ) ;

max_batch = MAXPOSNUM / (gradlen * sizeof(float)) ; // Memory allocation limit

if (max_batch > 65535)                     // Grid dimension limit
  max_batch = 65535 ;

if (max_batch > nc)                        // Must not have a batch with zero cases!
  max_batch = nc ;

n_subsets = nc / max_batch + 1 ;           // The +1 is required (integer truncation)

if (n_subsets < TrainParams.n_subsets)     // Allow user increase to prevent timeout
  n_subsets = TrainParams.n_subsets ;

if (n_subsets > nc) {  // Happens if user specifies a huge model for a tiny dataset
  ... Issue error message and abort
  }
```

If the CUDA initialization routine (presented soon) has not yet been called for this model architecture, which is flagged by a global variable, then we need to call it now. For it to know how much memory to allocate for casewise storage, we recompute the actual value of the maximum batch size that will be encountered when processing is done later. Then, if the device does not have the current set of model weights (again indicated by a global variable), we send the weights to the device.

We'll see the initialization parameters soon, but classifier && (target != NULL) rates special explanation. This tells whether the model being trained is a classifier. An autoencoder, even if destined to be part of a classifier, is itself *not* a classifier.

```
if (! cpx_cuda_initialized) {          // This global tells if model is initialized on device

  n_done = 0 ;
  for (ibatch=0 ; ibatch<n_subsets ; ibatch++) {
    n_in_batch = (nc - n_done) / (n_subsets - ibatch) ; // Cases left / batches left
    if (ibatch == 0 || n_in_batch > max_batch)
      max_batch = n_in_batch ;   // Keep track of maximum batch size
    n_done += n_in_batch ;
    }

  cpx_cuda_init ( is_complex , classifier && (target != NULL) , class_ids ,
                  nc , nin , max_neurons , input , nout ,
                  target , max_batch , n_layers , nhid , msg ) ;

  cpx_cuda_initialized = 1 ;
  }

if (cuda_weights_changed) {        // This global tells if device weights are current
  cuda_cpx_weights_to_device ( nin , nout , n_layers , nhid ,
                               weights , last_layer_weights ) ;
  cuda_weights_changed = 0 ;
  }
```

We initialize to zero the vector in which the gradient will be summed across all cases. Then the training set is broken into batches, each processed by the device separately.

```
for (i=0 ; i<n_weights ; i++)
  grad[i] = 0.0 ;

istart = 0 ;       // Batch start = training data start
n_done = 0 ;   // Number of training cases done in this epoch so far

for (ibatch=0 ; ibatch<n_subsets ; ibatch++) {
  n_in_batch = (nc - n_done) / (n_subsets - ibatch) ; // Cases left / batches left
  istop = istart + n_in_batch ;                        // Stop just before this index
```

First, the forward pass computes all activations. If this is a complex-domain model, the partial derivatives of the activation function will also be computed and saved. If this is a classification model being trained with supervision, SoftMax modification of the outputs is done.

```
for (ilayer=0 ; ilayer<n_layers-1 ; ilayer++)
   cuda_cpx_hidden_activation ( istart , istop , nhid[ilayer] , ilayer , 1 ) ;

cuda_cpx_output_activation ( istart , istop , nout ) ;

if (classifier && (target != NULL))
   cuda_cpx_softmax ( istart , istop ) ;
```

The backward pass computes the output delta and then all hidden-layer gradients except the first layer's (which receives the inputs). The first hidden layer's gradient is computed last of all. The routine cuda_cpx_fetch_gradient() sums the computed individual case gradients into the local gradient vector.

```
cuda_cpx_output_delta ( istart , istop , classifier && (target != NULL) , nout ) ;
cuda_cpx_output_gradient ( n_in_batch , nhid[n_layers-2] , n_layers-2 , nout ) ;

for (ilayer=n_layers-2 ; ilayer>0 ; ilayer--)
   cuda_cpx_subsequent_hidden_gradient ( n_in_batch , ilayer ,
               nhid[ilayer] , nhid[ilayer-1] , ilayer==n_layers-2 ) ;

cuda_cpx_first_hidden_gradient ( istart , istop , nin , nhid[0] , n_layers==2 ) ;

cuda_cpx_fetch_gradient ( n_in_batch , grad ) ;

n_done += n_in_batch ;
istart = istop ;
} // For all batches
```

After all batches have been processed, the summed gradient is divided by the number of cases and the number of outputs. If this is a real model or any type of model being trained supervised (in which case the imaginary parts of a complex prediction are ignored), this is correct. For a complex-domain autoencoder, we may also want to additionally divide by two, because the imaginary parts go into the error measure sum.

Then, if this model is a classifier being trained supervised, we compute the log-likelihood performance criterion. Otherwise, we compute the mean-squared-error criterion.

```
for (i=0 ; i<n_weights ; i++)
  grad[i] /= nc * nout ;

if (classifier && (target != NULL)) {
  cuda_cpx_ll ( nc , &mse ) ;          // Put log likelihood in mse
  mse /= nout ;                        // cuda_cpx_ll() divided by nc but not nout
  }

else
  ret_val = cuda_cpx_mse ( nc * nout , &mse ) ;
```

The final step is to compute the weight penalty and compensate the gradient for it. This is discussed in Volume I, so here I will just present the code. The first block handles the hidden-layer weights, and the second block handles the output layer.

```
wpen = TrainParams.wpen / n_weights ;
penalty = 0.0 ;

nin_this_layer = nin ;

for (ilayer=0 ; ilayer<n_layers-1 ; ilayer++) { // Do all hidden layers

  for (ineuron=0 ; ineuron<nhid[ilayer] ; ineuron++) {
    wptr = weights[ilayer] + ineuron * mult * (nin_this_layer+1) ;
    gptr = grad_ptr[ilayer] + ineuron * mult * (nin_this_layer+1) ;
    for (ivar=0 ; ivar<mult*nin_this_layer ; ivar++) {        // Do not include bias
      penalty += wptr[ivar] * wptr[ivar] ;
      gptr[ivar] -= 2.0 * wpen * wptr[ivar] ;
      }
    }
  nin_this_layer = nhid[ilayer] ;
  }
  for (ineuron=0 ; ineuron<nout ; ineuron++) {
    wptr = last_layer_weights + ineuron * n_last_layer_weights ;
    gptr = grad_ptr[n_layers-1] + ineuron * n_last_layer_weights ;
    for (ivar=0 ; ivar<mult*nin_this_layer ; ivar++) {       // Do not include bias
      penalty += wptr[ivar] * wptr[ivar] ;
```

```
         gptr[ivar] -= 2.0 * wpen * wptr[ivar] ;
         }
      }

   penalty *= wpen ;
   return mse + penalty ;
}
```

Device Initialization

The subroutine that initializes the CUDA device according to the characteristics of the model being trained is quite large, too unwieldy to print in its entirety here. It is available for free download from my web site. However, key variables, along with selected parts of the initialization algorithms, are presented. Here are variables that are declared at the top of the routine. Sometimes their comments will include *(complex)* or *(actual)*. The former means that for complex-domain models, the stored quantity refers to complex numbers (two actual numbers each), while the latter means that they are actual counts of numbers. This listing of variables is a handy reference for the individual routines that follow.

Also note that in a great many cases two variables will have the same name, except that one will begin with *d_* and the other with *h_*. The former variables are in the device's namespace, while the latter are in the host's namespace. This lets us cut way down on the number of parameters that must be passed in a kernel launch. At initialization time, the *d_* quantities are written to the device, meaning that they will be available to all kernel routines at runtime, without having to be passed as parameters.

```
// This is used as intermediary between device's float and hosts double
static float *fdata = NULL ;

static int n_out_weights ; // Total number of output weights (includes end padding)
static int n_hid_weights ; // Total number of hidden weights (includes end padding)

// These are for the reductions used in device_cpx_mse
// The number of threads MUST be a power of two!
// The number of blocks given here is a maximum. The actual number may be less.

#define REDUC_THREADS 256
#define REDUC_BLOCKS 64
```

```
static float *reduc_fdata = NULL ;

static int is_complex ;                  // Is this a complex model?
static int mult ;                        // 2 if complex, else 1
__constant__ int *d_is_classifier ;      // Is this a classifier?
__constant__ int d_complex ;             // Is this a complex model?
__constant__ int d_ncases ;              // Number of cases in complete training set
__constant__ int d_n_trn_inputs ;        // Number of first-layer inputs (complex)
__constant__ int d_ntarg ;               // Number of targets (output neurons) (complex)
__constant__ int d_ntarg_cols ;          // Ditto, extended to multiple of 128 bytes (actual)
__constant__ int d_n_layers ;            // Number of layers
__constant__ int d_mult ;                // 1 if real model, 2 if complex
__constant__ int d_autoencode ;          // If nonzero, include imaginary part in error

static    int *h_nhid = NULL ;           // Number of neurons in each hidden layer (complex)
__constant__ int *d_nhid ;
static    int *h_nhid_cols = NULL ;      // Ditto, extended to multiple of 128 bytes (actual)
__constant__ int *d_nhid_cols ;

static    float *h_trn_data = NULL ; // Raw training data; nc ases by mult*n_trn_inputs
__constant__ float *d_trn_data ;

static    float *h_targets = NULL ;      // Target data; ncases by ntarg (always strictly real)
__constant__ float *d_targets ;

static    int *h_class = NULL ;          // If classification (SoftMax), class id is here
__constant__ int *d_class ;

static    float *hidden_weights = NULL ; // Weight matricies for hidden layers
static    float **h_whid = NULL ;
__constant__ float **d_whid ;

static    float *h_wout = NULL ;             // Weight matrix for output layer, transpose of Host
__constant__ float *d_wout ;

static    double *activations = NULL ; // Activations of this layer, which we compute
static    double **h_act = NULL ;          // Array of pointers to each layer
__constant__ double **d_act ;
```

```
static     double *derivs = NULL ;       // Activation derivatives of this layer
static     double **h_drr = NULL ;       // Array of pointers to each layer for real/real
__constant__ double **d_drr ;
static     double **h_dii = NULL ;       // Array of pointers to each layer for imag/imag
__constant__ double **d_dii ;
static     double **h_dri = NULL ;       // Array of pointers to each layer for real/imaginary
__constant__ double **d_dri ;

static     double *h_output = NULL ; // Output activations, complex if complex model
__constant__ double *d_output ;

static     float *h_mse_out = NULL ; // For outputting performance measure
__constant__ float *d_mse_out ;

static     double *h_this_delta = NULL ; // Delta for current layer, complex if cpx model
__constant__ double *d_this_delta ;

static     double *h_prior_delta = NULL ;// Delta for nex t layer back
__constant__ double *d_prior_delta ;

// WARNING... If gradient is ever double instead of float, revise integer overflow check!

static     int h_gradlen ;               // Length of complete gradient for a case (actual)
__constant__ int d_gradlen ;
static     float *h_gradient = NULL ;   // Gradient for all layers, including output
__constant__ float *d_gradient ;
static     float **h_grad_ptr = NULL ; // Pointers to locations in grad for each layer
__constant__ float **d_grad_ptr ;
```

Here is the calling parameter list for the initialization routine. This is called before any of the other CUDA routines are invoked. It writes copies of essential parameters to the device so that they do not need to be passed as parameters, and it allocates memory.

```
int cpx_cuda_init (
  int complex ,      // Is this a complex-domain model?
  int classifier ,   // Is this for classification? (SoftMax outputs)
  int *class_ids ,   // Class ids if classifier
  int ncases ,       // Number of training cases
  int n_inputs ,     // Number of inputs (complex)
```

```
int ncols ,          // Number of columns in data matrix
double *data ,   // Input data, ncases rows by ncols columns; first n_inputs are used
int ntarg ,          // Number of targets (outputs; classes in classification) (complex)
double *targets ,// Targets, ncases by ntarg; always real, even for complex models
int max_batch , // Max size of any batch
int n_layers ,     // Number of layers of neurons, including output
int *nhid ,          // Number of neurons in each hidden layer (complex)
char *error_msg // Returns text of error if problem
)
```

Here are a few examples of how important constants are written to the device. We will find it handy to have mult be 2 for complex-domain models and 1 for real-domain models. The flag d_autoencode will be true if the targets are the model inputs, which the caller indicates by setting targets to NULL. Finally, all weights are aligned in memory so that rows of the matrix always start on an address that is a multiple of 128 bytes. This is done by padding the end of rows as necessary, and it can significantly speed transfer of weights from global memory through the cache. Here we show how the row length for the output weight matrix is bumped up to a multiple of 128 bytes (32 4-byte floats) for each part.

```
mult = complex ? 2 : 1 ;

cudaMemcpyToSymbol ( d_is_classifier , &classifier , sizeof(int) , 0 ,
                                  cudaMemcpyHostToDevice ) ;

k = (targets == NULL) ? 1 : 0 ;
cudaMemcpyToSymbol ( d_autoencode , &k , sizeof(int) , 0 ,
                                  cudaMemcpyHostToDevice ) ;

ntarg_cols = mult * ((ntarg + 31) / 32 * 32) ; // For alignment of weights to 128 bytes
```

We copy the number of neurons in each hidden layer to the device and also copy these values bumped up to 128-byte multiples. These save-to-device operations are done with three calls to CUDA functions.

1) Allocate device memory for the array of neuron counts.

2) Copy the counts from the host to the device.

3) Copy the device address of this array to a symbol on the device so it does not have to be passed as a parameter in a kernel launch.

```
memsize = (n_layers-1) * sizeof(int) ;
total_memory += memsize ;                    // This is just for user information

error_id = cudaMalloc ( (void **) &h_nhid , (size_t) memsize ) ;

error_id = cudaMemcpy ( h_nhid , nhid , (n_layers-1) * sizeof(int) ,
                        cudaMemcpyHostToDevice ) ;

error_id = cudaMemcpyToSymbol ( d_nhid , &h_nhid , sizeof(int *) , 0 ,
                                cudaMemcpyHostToDevice ) ;

for (i=0 ; i<n_layers-1 ; i++)          // Pad hidden-layer weight rows to 128 bytes
  nhid_cols[i] = mult * ((nhid[i] + 31) / 32 * 32) ;

memsize = (n_layers-1) * sizeof(int) ;
total_memory += memsize ;

error_id = cudaMalloc ( (void **) &h_nhid_cols , (size_t) memsize ) ;

error_id = cudaMemcpy ( h_nhid_cols , nhid_cols , (n_layers-1) * sizeof(int) ,
                        cudaMemcpyHostToDevice ) ;

error_id = cudaMemcpyToSymbol ( d_nhid_cols , &h_nhid_cols , sizeof(int *) , 0 ,
                                cudaMemcpyHostToDevice ) ;
```

The last bit of initialization we'll examine here is allocation of space for the hidden-layer weight matrices. We've studied less than one-tenth of the initialization code, but if you understand these topics, the rest of the code should be reasonably clear. The first step is to compute the total number of weights to be stored. We must include any end-of-row padding that is used to bring rows up to a multiple of 128 bytes. The +1 in this code is to accommodate the bias term. It's tempting to think that for a complex-domain model this should be +2. But remember that nhid_cols already incorporates a multiplier of 2 for complex models.

Also, we save this total count to the static variable n_hid_weights, which will be needed by the routine that copies weights from the host to the device.

```
n_total = 0 ;
n_prior = n_inputs ;
for (i=0 ; i<n_layers-1 ; i++) {
  n_total += nhid_cols[i] * (n_prior + 1) ; // Columns times rows in this layer
  n_prior = nhid[i] ;                        // Mult is included in nhid_cols, which is actual
  }

n_hid_weights = n_total ; // Needed in cuda_cpx_weights_to_device()
```

We need to allocate two blocks of memory on the device. The large block holds the weights (along with any end-of-row padding). A smaller block is an array of pointers to the addresses within the first block of the weight matrix for each hidden layer. We copy the address of this pointer array to a constant on the device so that it does not have to be passed as a parameter when the kernel is launched.

```
memsize = n_total * sizeof(float) ; // The weights
total_memory += memsize ;
error_id = cudaMalloc ( (void **) &hidden_weights , (size_t) memsize ) ;

memsize = (n_layers-1) * sizeof(float *) ; // Pointers to the weight matrix for each layer
total_memory += memsize ;
error_id = cudaMalloc ( (void **) &h_whid , (size_t) memsize ) ;

cudaMemcpyToSymbol ( d_whid , &h_whid , sizeof(void *) , 0 ,
                     cudaMemcpyHostToDevice ) ;
```

Finally, we compute the address within hidden_weights (on the device) of the weight matrix for each layer. Copy this array of pointers to the device.

```
float *fptr[MAX_LAYERS] ;

n_total = 0 ;
n_prior = n_inputs ;
for (i=0 ; i<n_layers-1 ; i++) {
  fptr[i] = hidden_weights + n_total ;
  n_total += nhid_cols[i] * (n_prior + 1) ;  // Columns times rows in this layer
  n_prior = nhid[i] ;                         // Mult is included in nhid_cols
  }

error_id = cudaMemcpy ( h_whid , &fptr[0] , (n_layers-1) * sizeof(float *) ,
                        cudaMemcpyHostToDevice ) ;
```

Copying Weights from Host to Device

The initialization routine briefly described in the prior section is called only once, so it does not copy any weights during initialization. This copying will be done often during training, each time the weights are updated by the training algorithm, so the copy process has its own routine. Here it is. A discussion is interspersed.

```
int cuda_cpx_weights_to_device (
  int n_inputs ,                   // Number of inputs, possibly complex
  int ntarg ,                      // Ditto outputs
  int n_layers ,                   // Number of layers, including output layer
  int *nhid ,                      // Number of hidden layers
  double **hid_weights ,           // Hidden layer weight matrices
  double *final_layer_weights )    // Output weight matrix
{
  int n_prior, ilayer, ineuron, ivar, ntarg_cols_each, nhid_cols_each ;
  double *wptr ;
  float *fptr ;
  char msg[256] ;
  cudaError_t error_id ;

  fptr = fdata ;
  n_prior = n_inputs ;

  for (ilayer=0 ; ilayer<n_layers-1 ; ilayer++) {
    wptr = hid_weights[ilayer] ;
    nhid_cols_each = (nhid[ilayer] + 31) / 32 * 32 ; // For memory alignment to 128 bytes

    for (ivar=0 ; ivar<=n_prior ; ivar++) {

      // Real part goes in first half of row for complex-domain models
      for (ineuron=0 ; ineuron<nhid[ilayer] ; ineuron++)
        *fptr++ = (float) wptr[mult*(ineuron*(n_prior+1)+ivar)] ;

      while (ineuron++ < nhid_cols_each)      // Fill in padding to multiple of 128 bytes
        *fptr++ = 0.0f ;
```

```
      // Imaginary part goes in second half of row
      if (is_complex) {
         for (ineuron=0 ; ineuron<nhid[ilay er] ; ineuron++)
            *fptr++ = (float) wptr[2*(ineuron*(n_prior+1)+iv ar)+1] ;
         while (ineuron++ < nhid_cols_each) // Fill in padding
            *fptr++ = 0.0f ;
         }
      } // For ivar

   n_prior = nhid[ilayer] ;
   }

   assert ( fptr == fdata + n_hid_weights ) ;

   error_id = cudaMemcpy ( hidden_weights , fdata , n_hid_weights * sizeof(float) ,
                           cudaMemcpyHostToDevice ) ;
```

Notice in the previous code that we transpose the weight matrix. The reason for this will become clear later when kernel function memory access is discussed. Also, on the host and for everything else here except weight matrices, complex numbers are stored with the imaginary part immediately following the real part. But weight matrices are stored with the real part in the first half of each row and the imaginary part in the second half. Each of these halves is padded to a multiple of 128 bytes. Again, the reason for this will become clear later.

The output weight matrix is handled similarly, and this code is shown on the following page.

```
   fptr = fdata ;
   wptr = final_layer_weights ;
   ntarg_cols_each = (ntarg + 31) / 32 * 32 ; // For memory alignment to 128 bytes

   for (ivar=0 ; ivar<=n_prior ; ivar++) { // Store as transpose

      // Real part goes in the first half of the row for a complex-domain model
      for (ineuron=0 ; ineuron<ntarg ; ineuron++)
         *fptr++ = (float) wptr[mult*(ineuron*(n_prior+1)+ivar)] ;

      while (ineuron++ < ntarg_cols_each)      // Fill in the padding to 128 byte multiple
         *fptr++ = 0.0f ;
```

```
    if (is_complex) {
      // Imaginary part
      for (ineuron=0 ; ineuron<ntarg ; ineuron++)
        *fptr++ = (float) wptr[2*(ineuron*(n_prior+1)+iv ar)+1] ;

      while (ineuron++ < ntarg_cols_each)     // Fill in padding
        *fptr++ = 0.0f ;
      }
    }

  assert ( fptr == fdata + n_out_weights ) ;

  error_id = cudaMemcpy ( h_wout , fdata , n_out_weights * sizeof(float) ,
                          cudaMemcpyHostToDevice ) ;

  return 0 ;
}
```

Activation and Its Derivatives

The source code available on my web site includes specialized routines for real and complex-domain models, with and without derivatives. Here we will discuss only the most complicated version, that for computing derivatives as well as activations for complex-domain models. All other versions are subsets of this one.

We begin with the host code for the algorithm. This is called from the training routine, and it in turn launches the CUDA kernel routines.

This is where some background in CUDA computation is required. Volume I made no assumptions about your knowledge of this subject, and it included much introductory material, with a special focus on optimal memory access. If you have no background in this area, you will need to read either Volume I or any of the excellent introductory CUDA books available.

Work will be distributed by assigning neurons in a layer to threads. As is my usual habit for neural networks, I impose a limit of four warps per block, a good compromise with wide safety regions on both sides.

```
int cuda_cpx_hidden_activation (
  int istart ,       // First case in this batch
  int istop ,        // One past last case
  int nhid ,         // Number of hidden neurons in this layer
  int ilayer ,       // Layer to process
  int need_deriv  // Also compute derivatives?
  )
{
  int warpsize, threads_per_block ;
  char msg[256] ;
  dim3 block_launch ;
  cudaError_t error_id ;

  warpsize = deviceProp.warpSize ;      // Threads per warp, likely 32 well into the future

  threads_per_block = (nhid + warpsize - 1) / warpsize * warpsize ;
  if (threads_per_block > 4 * warpsize)
    threads_per_block = 4 * warpsize ;
```

The launch is two-dimensional, with the *x* coordinate identifying the neuron whose activation is being computed and the *y* coordinate identifying the case being processed. We choose which of three kernels to use depending on whether it is a real or complex-domain model and whether we need derivatives. Derivatives are needed for gradient descent but not needed for simulated annealing based on performance. Since derivatives add considerable computational expense, we should not compute them unless they are needed. And the derivatives for a real-domain model are so simple that we don't bother precomputing and saving them; they are computed on the fly during training.

```
  block_launch.x = (nhid + threads_per_block - 1) / threads_per_block ;
  block_launch.y = istop - istart ;
  block_launch.z = 1 ;

  if (is_complex && need_deriv)
    device_cpx_hidden_activation_d <<< block_launch , threads_per_block >>>
                                                  ( istart , istop , ilayer ) ;
  else if (is_complex)
    device_cpx_hidden_activation_c <<< block_launch , threads_per_block >>>
                                                  ( istart , istop , ilayer ) ;
```

```
  else
    device_cpx_hidden_activation_r <<< block_launch , threads_per_block >>>
                                            ( istart , istop , ilayer ) ;

  cudaDeviceSynchronize() ;
  error_id = cudaGetLastError () ;

  ... Handle errors ...

  return 0 ;
}
```

The kernel function begins by getting the thread identifier (hidden neuron here) and immediately returning if it is beyond the number of neurons in this layer. The variable nhid_cols is the number of columns in the weight matrix for this layer. It includes however many padding columns are needed to bring the dimension of the real and imaginary row parts each up to a multiple of 128 bytes. Recall that memory fetches from the cache are 128 bytes at a time and always start at an address that is a multiple of 128 bytes. We also fetch the case index.

```
__global__ void device_cpx_hidden_activation_d (
  int istart ,    // First case in this batch
  int istop ,     // One past last case
  int ilayer      // Layer to process
  )
{
  int k, icase, ihid, i_input, n_inputs, nhid_cols ;
  float *f_inptr, *wptr ;
  double rsum, isum, *d_inptr, len_sq, raw_length, squashed_length, ratio ;
  double *actptr, *drrptr, *diiptr, *driptr, deriv, temp ;

  ihid = blockIdx.x * blockDim.x + threadIdx.x ;

  if (ihid >= d_nhid[ilayer])
    return ;

  nhid_cols = d_nhid_cols[ilayer] ;  // Actual, a multiple of 128 bytes (32 floats)
  k = nhid_cols / 2 ;                // Real and imaginary parts of weights separated by this

  icase = blockIdx.y ;
```

We use wptr to point to the first element of the weight vector for this hidden neuron. Recall that the weight matrix is stored as the transpose of the copy on the host, so the hidden neuron in the current layer changes fastest. Also get pointers to the activation and derivative vectors that will be computed in this routine.

```
wptr = d_whid[ilayer] + ihid ;
actptr = d_act[ilayer] + 2 * (icase * d_nhid[ilayer] + ihid) ;
drrptr = d_drr[ilayer] ;
diiptr = d_dii[ilayer] ;
driptr = d_dri[ilayer] ;
```

The activation is computed in one of two blocks, depending on whether this is the first hidden layer (fed by the model inputs) or a subsequent hidden layer (fed by the prior hidden layer). Model inputs are floats, while activations are doubles, so we need different pointers. The net input to the neuron is computed by application of Equation 4-5.

It should now be clear why the weight matrices are stored the way they are. The input pointers in this expensive loop do not depend on the hidden neuron being computed, so they are identical for all threads in a warp. Only wptr depends on the thread ID. Because its first value is guaranteed to be a multiple of 128 bytes and we add nhid_cols in the loop, the first thread in each warp accesses memory at this magic address, and subsequent threads access contiguous locations. The imaginary parts are offset by a multiple of 128 bytes, so they, too, are perfectly aligned.

```
rsum = isum = 0.0 ;
if (ilayer == 0) {                                     // First hidden layer
  n_inputs = d_n_trn_inputs ;                          // This many complex inputs
  f_inptr = d_trn_data + (icase+istart)*2*n_inputs ;   // Inputs are here
  for (i_input=0 ; i_input<n_inputs ; i_input++) {     // For all of them
    rsum += *wptr * *f_inptr - *(wptr+k) * *(f_inptr+1) ; // Equation 4-5
    isum += *wptr * *(f_inptr+1) + *(wptr+k) * *f_inptr ; // Ditto
    wptr += nhid_cols ;                                // Next row (input)
    f_inptr += 2 ;                                     // Inputs stored (real, imagainry)
    }
  rsum += *wptr ;                                      // Bias
  isum += *(wptr+k) ;
  }
```

```
else {                                                  // Subsequent hidden layer
  n_inputs = d_nhid[ilayer-1] ;                         // Size of prior layer
  d_inptr = d_act[ilayer-1] + icase*2*n_inputs ;        // Activation of prior layer
  for (i_input=0 ; i_input<n_inputs ; i_input++) {      // For all prior-layer neurons
    rsum += *wptr * *d_inptr - *(wptr+k) * *(d_inptr+1) ;
    isum += *wptr * *(d_inptr+1) + *(wptr+k) * *d_inptr ;
    wptr += nhid_cols ;
    d_inptr += 2 ;
    }
  rsum += *wptr ; // Bias
  isum += *(wptr+k) ;
  }
```

It might be worth a bit more examination of memory alignment. When a running kernel program requests something from global memory, the hardware fetches it in blocks of 128 bytes aligned on 128-byte boundaries, even if just a single byte is needed. We must make maximum use of it!

The weight matrix d_whid[ilayer] is guaranteed to start on a 128-byte boundary. The first thread in the first block has ihid=0, so in the first pass of the i_input loop, 32 floats (128 bytes) are fetched, with the first float used by the first thread. The second thread has ihid=1, so it uses the second float, and so on. Thus, a single 128-byte fetch gets *wptr for all 32 threads in the warp! And *(wptr+k) is offset by a multiple of 128 bytes, so the same holds for it. Finally, each pass through the loop adds a multiple of 128 bytes to wptr, preserving this perfect alignment. Yes!

We now use Equation 4-20 to compute the activation.

```
len_sq = rsum * rsum + isum * isum + 1.e-60 ;
raw_length = sqrt ( len_sq ) ;
squashed_length = tanh ( 1.5 * raw_length ) ;
ratio = squashed_length / raw_length ;

*actptr = rsum * ratio ;
*(actptr+1) = isum * ratio ;
```

Finally, we use the algorithms described on page 123 to compute the partial derivatives of the activation function. At first glance, the correspondence between the lines of code shown next and the equations of that section may not be clear.

This is because the mathematical operations have been optimized for speed and accuracy. A minute or two with pencil and paper will confirm the correctness of this implementation.

```
   deriv = 1.5 * (1.0 - squashed_length * squashed_length) ;
   temp = (deriv - ratio) / len_sq ;

   k = icase * d_nhid[ilayer] + ihid ;
   drrptr[k] = ratio + rsum * rsum * temp ;
   diiptr[k] = ratio + isum * isum * temp ;
   driptr[k] = rsum * isum * temp ;
}
```

Output Activation

Computation of the output activation is a subset of the hidden activation just shown because the activation function is linear and its partial derivatives are not needed. We'll dispense with the mundane launch code and jump right to the kernel routine.

In my program design, the outputs are preserved for all cases, while intermediate items such as activations, derivatives, and computed gradients are preserved only for the cases within a batch. This per-batch structure saves a tremendous amount of precious on-board memory, while having all outputs available on completion speeds computation of performance measures, and also lays the groundwork for more advanced operations that may be added later. So, because we are keeping all outputs as they are computed, we need to pass to the kernel the starting case index of this batch, istart.

Work is assigned by mapping outputs to threads. This is inefficient if there are few targets because warps will contain potentially numerous threads that are never executed. On the other hand, computing outputs almost always takes up a tiny percentage of the total execution time, so inefficiency here is of little or no consequence. Moreover, when autoencoding layers or groups of layers are trained, we will have (in virtually all practical applications) an enormous number of targets, meaning that in the situations where it matters, efficiency will not suffer.

```
__global__ void device_cpx_output_activation_c (
   int istart      // First case in this batch
   )
```

```
{
  int k, icase, iout, n_inputs, i_input, ilayer ;
  double rsum, isum, *inptr ;
  float *wptr ;

  iout = blockIdx.x * blockDim.x + threadIdx.x ;        // Thread index identifies output

  if (iout >= d_ntarg)                                   // Is this thread beyond outputs?
    return ;
```

Get the index and size of the last hidden layer, which feeds the output layer. The current implementation does not use CUDA for models that have no hidden layer, so we don't have to worry about direct input-to-output, although some code is ready for this addition if it ever happens. Also, get the case number (within the batch, not within the complete training set) from the block index. Set wptr to the address of the first weight for this output neuron. Set inptr to the activation vector of the hidden layer and the case that feeds the output layer. Set k to the distance within each row of the weight matrix that separates the imaginary part of each weight from the real part.

```
  ilayer = d_n_layers - 2 ;
  n_inputs = d_nhid[ilayer] ;

  icase = blockIdx.y ;
  wptr = d_wout + iout ;

  inptr = d_act[ilayer] + icase * 2 * n_inputs ;
  rsum = isum = 0.0 ;
  k = d_ntarg_cols / 2 ; // Real and imaginary parts of weights separated by this
```

The activations are computed with a straightforward application of Equation 4-5. Perfect memory alignment of the weight matrix is obtained in the same way as for the hidden-layer weights. The computed outputs do not have perfect alignment; they are strided by the multiplier of two. But this is an insignificant fraction of the total time.

```
  for (i_input=0 ; i_input<n_inputs ; i_input++) { // wout is transpose of Host
    rsum += *wptr * *inptr - *(wptr+k) * *(inptr+1) ;
    isum += *wptr * *(inptr+1) + *(wptr+k) * *inptr ;
    wptr += d_ntarg_cols ;
    inptr += 2 ;
    }
```

```
   rsum += *wptr ; // Bias
   isum += *(wptr+k) ;

   k = 2*((icase+istart)*d_ntarg+iout) ;
   d_output[k] = rsum ;
   d_output[k+1] = isum ;
}
```

SoftMax Modification of Outputs

If this model is a classifier and we are not doing autoencoding in the current training operation, we must modify the outputs as shown in Equation 4-34 and Equation 4-35. Because this operation involves summing across outputs, we must assign threads to cases rather than outputs.

We limit the logit to 300 to prevent rare but disastrous exponential overflows. Memory access to the output vector is very inefficient because the thread/case index causes it to stride severely. However, this routine is extremely fast relative to the other kernels, so inefficiency is totally innocuous. Also, exponentiation hides memory waits well.

```
__global__ void device_cpx_softmax (
   int istart ,      // First case in this batch
   int istop         // One past last case
   )
{
   int icase, iout ;
   double *outptr, sum ;

   icase = blockIdx.x * blockDim.x + threadIdx.x ;

   if (icase >= istop - istart)
      return ;

   outptr = d_output + (icase + istart) * d_mult * d_ntarg ; // Output vector for this case
   sum = 0.0 ;

   for (iout=0 ; iout<d_ntarg ; iout++) { // Imaginary part plays no role
      if (outptr[d_mult*iout] < 300.0)
         outptr[d_mult*iout] = __expf ( outptr[d_mult*iout] ) ;
```

```
      else
         outptr[d_mult*iout] = __expf ( 300.0 ) ;
      sum += outptr[d_mult*iout] ;
      }

   for (iout=0 ; iout<d_ntarg ; iout++)
      outptr[d_mult*iout] /= sum ;
}
```

Output Delta

Computation of the output delta (Equation 4-33, a simplification of Equation 4-32) is a little fussy because we may or may not be autoencoding and the model may be real or complex-domain. SoftMax output deltas will be presented in the next section. Once again, the thread index identifies the output neuron, and the *y* coordinate of the block index is the case.

```
__global__ void device_cpx_output_delta (
   int istart ,    // First case in this batch
   int istop ,     // One past last case
   int ntarg       // Number of targets (outputs)
   )
{
   int j, k, icase, iout ;

   iout = blockIdx.x * blockDim.x + threadIdx.x ;
   if (iout >= d_ntarg)
      return ;

   icase = blockIdx.y ;
```

If we are autoencoding, d_targets (set during initialization) points to the inputs, which are full complex if this is a complex model. In all other cases, the targets are strictly real.

```
  if (d_autoencode) {
    if (d_complex) {
      j = 2 * (icase * ntarg + iout ) ;              // Delta is indexed per batch
      k = 2 * ((icase + istart) * ntarg + iout) ;  // Outputs and target indexed in entire data
      d_this_delta[j] = 2.0 * (d_targets[k] - d_output[k]) ;
      d_this_delta[j+1] = 2.0 * (d_targets[k+1] - d_output[k+1]) ;
      }
    else {
      k = (icase + istart) * ntarg + iout ;
      d_this_delta[icase*ntarg+iout] = 2.0 * (d_targets[k] - d_output[k]) ;
      }
    }
```

If we are not autoencoding, d_targets is strictly real, and we ignore the imaginary part of predictions.

```
  else {
    j = d_mult * (icase * ntarg + iout) ;        // d_mult is 2 for complex, 1 for real
    k = (icase+istart)*ntarg+iout ;
    d_this_delta[j] = 2.0 * (d_targets[k] - d_output[d_mult*k]) ;
    if (d_complex)
      d_this_delta[j+1] = 0.0 ;                  // Imaginary part of output is ignored
    }
}
```

Delta for SoftMax Outputs

Computing delta for SoftMax outputs is easier because we know that the model currently being trained is not an autoencoder; we are training a complete model to predict class targets. It may still be a complex-domain model, in which case we ignore the imaginary part of the predictions. But regardless, all we need is a simple implementation of Equation 4-37.

```
__global__ void device_cpx_softmax_delta (
  int istart ,    // First case in this batch
  int istop ,     // One past last case
  int ntarg       // Number of targets (outputs)
  )
```

```
{
  int icase, iout ;

  iout = blockIdx.x * blockDim.x + threadIdx.x ;

  if (iout >= d_ntarg)
    return ;

  icase = blockIdx.y ;

  d_this_delta[d_mult*(icase*ntarg+iout)] = ( ((iout == d_class[icase+istart]) ? 1.0 : 0.0)
                                     - d_output[d_m ult*((icase+istart)*ntarg+iout)]) ;
  if (d_complex)
    d_this_delta[2*(icase*ntarg+iout)+1] = 0.0 ;        // Ignore imaginary part of output
}
```

Output Gradient

Going from the output delta to the output gradient is relatively easy because we are dealing with the last layer in the network; there are no additional layers through which errors are impacted. All we need to do is use Equation 4-31 to evaluate Equation 4-29. The gradient is the partial derivative of an output error with respect to the weight leading from a hidden neuron, evaluated for a case. So, the launch needs to be three-dimensional. The *x* dimension (thread) specifies the hidden neuron, *y* specifies the case, and *z* is the output neuron. We get these three quantities as well as the size of the hidden layer.

```
__global__ void device_cpx_output_gradient_c (
  int nc , // Number of cases in batch
  int ilayer // Hidden layer which feeds the output layer
  )
{
  int k, icase, iout, ihid, nhid ;
  float *gptr ;
  double r_delta, i_delta, r_prev, i_prev ;

  ihid = blockIdx.x * blockDim.x + threadIdx.x ;
  nhid = d_nhid[ilayer] ;      // Neurons in last hidden layer
  icase = blockIdx.y ;
```

If the thread exceeds the number of hidden neurons, we return. Usually the thread will identify a hidden neuron, in which case we get its activation. Exactly once the "activation" is 1.0+0*i*, the bias.

```
if (ihid > nhid)
   return ;
else if (ihid < nhid) {              // Usual situation: get this neuron's activation
   k = 2 * (icase * nhid + ihid) ;
   r_prev = d_act[ilayer][k] ;
   i_prev = d_act[ilayer][k+1] ;
   }
else {                               // When ihid == nhid, process the bias
   r_prev = 1.0 ;                    // Bias is 1.0 + 0 i
   i_prev = 0.0 ;
   }
```

Get delta for this output and point to the output gradient in the grand gradient vector. Then evaluate the basic gradient equations. Memory access to delta and the gradient is inefficiently strided, but this routine is such a tiny fraction of the total run time that inefficiency is of no consequence.

```
   iout = blockIdx.z ;

   k = 2 * (icase * d_ntarg + iout) ;
   r_delta = d_this_delta[k] ;
   i_delta = d_this_delta[k+1] ;

   gptr = d_grad_ptr[ilayer+1] + icase * d_gradlen ; // Gradient of output layer

   k = 2 * (iout * (nhid + 1) + ihid) ;
   gptr[k] = r_delta * r_prev + i_delta * i_prev ; // Eq (4-31) on Page 129 for Eq (4-29)
   gptr[k+1] = -r_delta * i_prev + i_delta * r_prev ;
}
```

Gradient of the First Hidden Layer

In most applications, the input layer is the largest. Moreover, the first hidden layer, by definition, feeds at least one more layer, meaning that the summation of Equation 4-41 (depicted in Figure 4-10) must be performed. Between these two facts, computing the gradient of the first hidden layer is nearly always the most ravenous eater of time. On my computers with several different nVidia display boards, this operation, combined with the gradient for any subsequent hidden layers, typically takes up around 95 percent of the total execution time, often even more. Thus, it must be tuned for peak efficiency.

By analogy with the output gradient, the launch is three-dimensional, with *x* being the feed, *y* being the case, and *z* being the destination neuron.

```
__global__ void device_cpx_first_hidden_gradient_c (
   int istart ,                    // First case in this batch
   int istop ,                     // One past last case
   int only_hidden                 // Is this the only hidden layer?
   )
{
   int j, k, icase, iin, ihid, nhid, n_next ;
   float *gptr, *next_weights ;
   double *delta_ptr, drr, dii, dri ;
   double r_delta, i_delta, r_prev, i_prev, rsum, isum ;

   iin = blockIdx.x * blockDim.x + threadIdx.x ;
   icase = blockIdx.y ;

   if (iin > d_n_trn_inputs)              // If thread is beyond all inputs, nothing to do
      return ;
   else if (iin < d_n_trn_inputs) {       // Normal situation: get this feed from input
      j = 2 * ((icase + istart) * d_n_trn_inputs + iin) ;
      r_prev = d_trn_data[j] ;
      i_prev = d_trn_data[j+1] ;
      }
   else {                                 // Just once we handle the bias
      r_prev = 1.0 ;
      i_prev = 0.0 ;
      }
```

Get the index of the neuron under consideration, as well as the size of this first hidden layer. The summation of Equation 4-41 will be across the layer following this one. If this is the only hidden layer in the model, the next layer is the output layer. Otherwise, it's the next hidden layer. We get n_next as the number of neurons in the next layer and get a pointer to the weights for this layer. Finally, set k to half the row length, which is the offset by which the imaginary parts of the weights are separated from the real parts.

```
ihid = blockIdx.z ;
nhid = d_nhid[0] ;            // Neurons in this hidden layer

if (only_hidden) {            // Next layer is output layer?
  n_next = d_ntarg ;
  next_weights = d_wout + ihid * d_ntarg_cols ;
  k = d_ntarg_cols / 2 ;    // Real and imaginary parts are separated by this
  }
else {
  n_next = d_nhid[1] ;      // This many neurons in next layer
  next_weights = d_whid[1] + ihid * d_nhid_cols[1] ; // Their weights start here
  k = d_nhid_cols[1] / 2 ;
  }
```

Get a pointer to the current delta, and then sum Equation 4-41. Especially if this is an autoencoding model or one with several large hidden layers, this loop consumes the vast majority of execution time. Observe that none of these memory fetches are related to the thread index, so the scheduler should be able to massively share fetched data among the threads in a warp. The implication is that the floating-point pipeline will be the limiting factor in the vast majority of large applications.

```
delta_ptr = d_this_delta + icase * 2 * n_next ;        // Delta for this case

rsum = isum = 0.0 ;
for (j=0 ; j<n_next ; j++) {                           // Equation 4-41
  rsum += delta_ptr[2*j] * next_weights[j] +
           delta_ptr[2*j+1] * next_weights[j+k] ;
  isum += -delta_ptr[2*j] * next_weights[j+k] +
            delta_ptr[2*j+1] * next_weights[j] ;
  }
```

The time-consuming part is done. We get the partial derivatives of the activation function for computing delta with Equation 4-40. Some programmers may want to avoid the temporary variables and insert the subscripted quantities directly. However, the CUDA compiler should place them in ultra-fast registers, meaning that there is no speed penalty, and the program is clearer when written this way.

```
    j = icase * nhid + ihid ;
    drr = d_drr[0][j] ;
    dii = d_dii[0][j] ;
    dri = d_dri[0][j] ;

    r_delta = rsum * drr + isum * dri ;           // Equation 4-40
    i_delta = rsum * dri + isum * dii ;
```

Last of all we use Equation 4-38 to compute the gradient. The saves to gptr are inefficiently strided, but the cost is negligible compared to the expense of the earlier summation loop.

```
    gptr = d_grad_ptr[0] + icase * d_gradlen ; // Gradient of first hidden layer
    j = 2 * (ihid * (d_n_trn_inputs + 1) + iin) ;
    gptr[j] =      r_delta * r_prev + i_delta * i_prev ;
    gptr[j+1] = - r_delta * i_prev + i_delta * r_prev ;
}
```

Gradient of a Subsequent Hidden Layer

Computation of the gradient for hidden layers after the first is practically identical to what was just presented, so there is no point in listing the code here. The differences are minor. Rather than hard-coding the layer numbers (0 for the current and 1 for the next hidden layer), we must pass a layer number as a launch parameter. Also, we must save the delta computed near the end of the routine in prior_delta so that it can be used when computing the gradient of the next layer back. Complete source code for these operations is available for free download from my web site.

Mean Squared Error

Recall that we saved the outputs for all training cases, and we have the targets for all cases. Computing the mean squared error is a simple matter of summing the squared differences between predictions and targets. The most efficient by far method for doing this is a standard parallel-processing algorithm called *reduction*.

Volume I of this series provided a detailed description of the reduction algorithm, and other descriptions are available in other texts. Also, it is quite complicated, so we will not repeat it here. If you are not familiar with reduction, you may well be baffled by the code listing, but brainstorming through it unaided is guaranteed to be a transformative experience.

There are no calling parameters, as it just processes the entire training set. There are d_ncases training cases and d_ntarg outputs. If the model is a complex-domain autoencoder, the imaginary parts of the outputs and targets are included in the error computation. But for supervised training of a complex-domain model, the imaginary parts are ignored. The kernel function begins as shown here:

```
__global__ void device_cpx_mse ()
{
  __shared__ double partial_mse[REDUC_THREADS] ;
  int i, index ;
  unsigned int n ;
  double diff, sum_mse ;

  index = threadIdx.x ;
  n = d_ncases * d_ntarg ;
  sum_mse = 0.0 ;
```

If we are autoencoding, d_targets points to the inputs, which are full complex if this is a complex-domain model. In all other cases, the targets are strictly real. Note that memory accesses are inefficiently strided here, but this is of no consequence because execution time for this algorithm is a tiny fraction of the total program time.

```
  if (d_autoencode) {

    if (d_complex) {
      for (i=blockIdx.x*blockDim.x+index ; i<n ; i+=blockDim.x*gridDim.x) {
        diff = d_output[2*i] - d_targets[2*i] ;             // Real part
        sum_mse += diff * diff ;
        diff = d_output[2*i+1] - d_targets[2*i+1] ;         // Imaginary part
        sum_mse += diff * diff ;
        }
      } // If complex-domain model
```

```
    else {      // Real-domain model
      for (i=blockIdx.x*blockDim.x+index ; i<n ; i+=blockDim.x*gridDim.x) {
        diff = d_output[i] - d_targets[i] ;
        sum_mse += diff * diff ;
        }
      } // Real-domain model
    } // If autoencoding
```

If we are not autoencoding, d_targets is strictly real, and we ignore the imaginary part (if complex) of predictions. This is why d_mult is used for indexing the outputs, but no multiplier is used for the targets.

```
  else {
    for (i=blockIdx.x*blockDim.x+index ; i<n ; i+=blockDim.x*gridDim.x) {
      diff = d_output[d_mult*i] - d_targets[i] ; // Imaginary part is ignored
      sum_mse += diff * diff ;
      }
    }
```

All that remains is to finish the reduction algorithm, as shown on the next page.

```
    partial_mse[index] = sum_mse ;
    __syncthreads() ;

    for (i=blockDim.x>>1 ; i ; i>>=1) {
      if (index < i)
        partial_mse[index] += partial_mse[index+i] ;
      __syncthreads() ;
      }

    if (index == 0)
      d_mse_out[blockIdx.x] = partial_mse[0] ;
}
```

For completeness and because reduction is such a cryptic algorithm for the uninitiated, here is the code for launching the kernel function just shown:

```
int cuda_cpx_mse (
  int n ,              // Number of values; ncases * ntarg
  double *mse      // Computed mse criterion
  )
{
  int i, blocks_per_grid ;
  double sum ;
  char msg[256] ;
  cudaError_t error_id ;

  blocks_per_grid = (n + REDUC_THREADS - 1) / REDUC_THREADS ;
  if (blocks_per_grid > REDUC_BLOCKS)
    blocks_per_grid = REDUC_BLOCKS ;

  device_cpx_mse <<< blocks_per_grid , REDUC_THREADS >>> () ;
  cudaDeviceSynchronize() ;

  error_id = cudaMemcpy ( reduc_fdata , h_mse_out , blocks_per_grid * sizeof(float) ,
                                       cudaMemcpyDeviceToHost ) ;
  sum = 0.0 ;
  for (i=0 ; i<blocks_per_grid ; i++)
    sum += reduc_fdata[i] ;
  *mse = sum / n ;
  return 0 ;
}
```

The Log Likelihood Criterion for Classification

If the outputs are SoftMax class probabilities, computation of the log likelihood optimization criterion is a lot easier than what we just saw for mean squared error. This is because we know the model cannot be an autoencoder, and if it is complex domain, the imaginary parts will always be ignored. Here is this code, presented without explanation since it is just a simplification of what was just seen. The criterion is given by Equation 4-36, and 1.e-30 prevents accidental floating-point disaster.

```
__global__ void device_cpx_ll ()
{
  __shared__ double partial_ll[REDUC_THREADS] ;
  int i, n, ntarg, index ;
  double sum_ll ;

  index = threadIdx.x ;
  n = d_ncases ;
  ntarg = d_ntarg ;

  sum_ll = 0.0 ;
  for (i=blockIdx.x*blockDim.x+index ; i<n ; i+=blockDim.x*gridDim.x)
    sum_ll -= log ( d_output[d_mult*(i*ntarg+d_class[i])] + 1.e-30 ) ;

  partial_ll[index] = sum_ll ;
  __syncthreads() ;

  for (i=blockDim.x>>1 ; i ; i>>=1) {
    if (index < i)
      partial_ll[index] += partial_ll[index+i] ;
    __syncthreads() ;
    }

  if (index == 0)
    d_mse_out[blockIdx.x] = partial_ll[0] ;
}
```

An Analysis

In practically all applications, the dominant consumer of device computation time is the routine that computes the gradient of the first hidden layer. Here is a brief analysis of its resource usage, aimed primarily at readers already familiar with CUDA programming. This is for training a 250 complex neuron autoencoder with MNIST data.

Figure 4-11 shows how work is distributed among the multiprocessors. It reveals that the work load is excellently balanced. Figure 4-12 shows that the arithmetic pipe (the right bar in the left graph) is running at nearly 100 percent capacity and hence is the limiting factor. This is evidence that the kernel is not wasting time waiting for inefficient memory transactions.

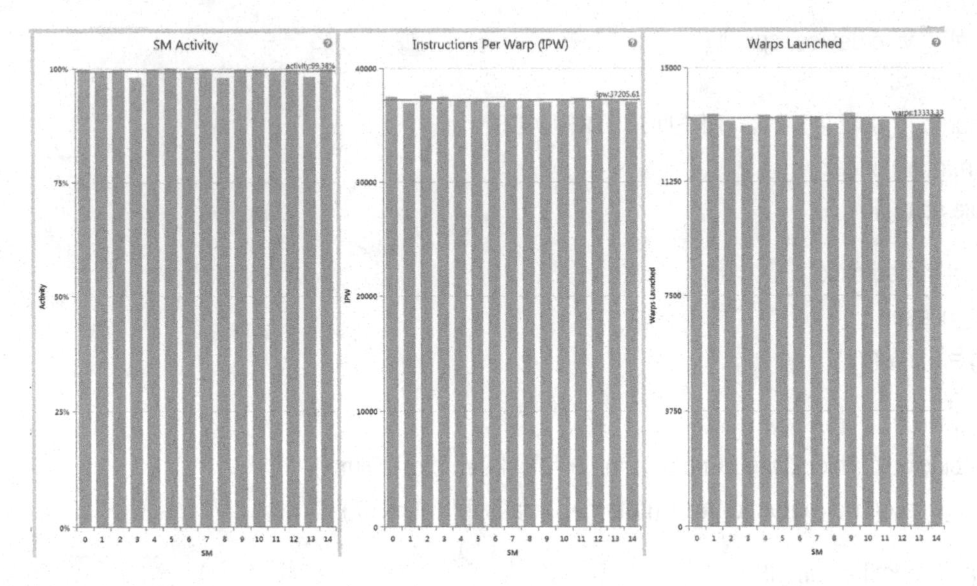

Figure 4-11. *Work distribution*

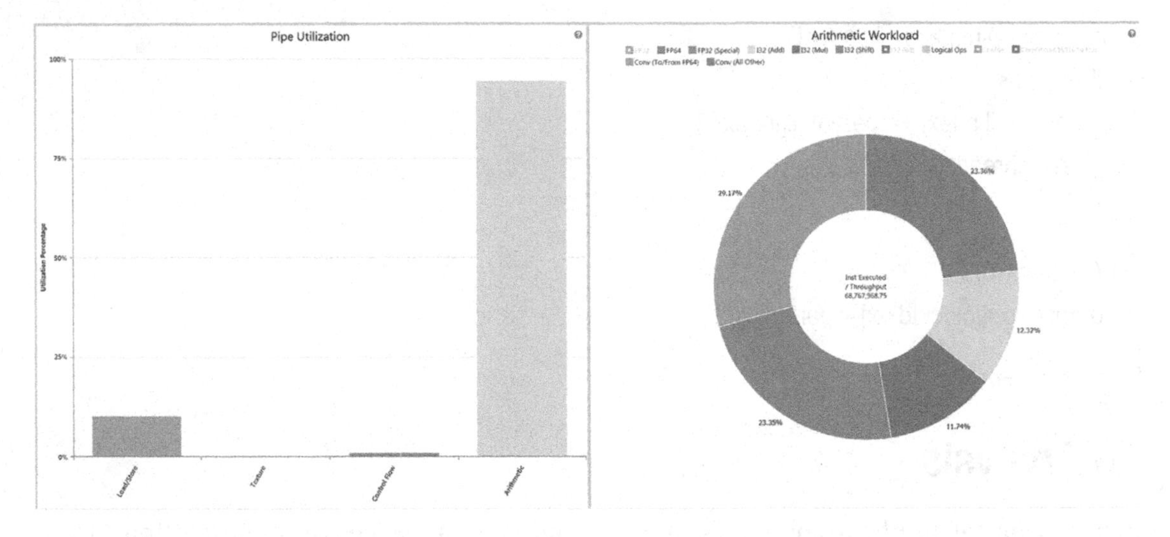

Figure 4-12. *Pipe utilization*

Perhaps the most interesting part of this analysis is the source code shown in Figure 4-13 and originally discussed on page 180. The second column of numbers, *Thread Instructions Executed,* is the number of low-level instructions executed by each thread for the corresponding line of code. If you look at the n_next loop about two-thirds of the way down, you'll see that the rsum and isum operations in this loop profoundly dominate execution in the kernel, so their efficiency is of paramount importance. The *L1 above ideal transactions* count is the number of memory transactions whose size exceeded the ideal size. Basically, such an operation transferred more data to or from the L1 cache than was needed by the kernel. We see that this count is actually negative! This reflects the fact that the warp scheduler was able to reuse for multiple threads data that was already fetched. We also see this in the *L1 transfer overhead* figure of 0.7, which is less than the "ideal" of exactly 1. The more expensive L2 overhead is even better at 0.2.

The worst memory access figures here are in the last two lines, in which the computed gradient is saved. A transfer overhead of 3.0 is fairly poor. It comes about because not only is the gradient not aligned to 128-byte addresses, but it depends on iin, the thread index, and this address is multiplied by 2 to accommodate the fact that the real and imaginary parts of the gradient are contiguous. This multiplication causes memory striding.

This would be a major deal killer if it weren't for two crucial facts. First, the number of times these inefficient saves are executed is a small, often minuscule, fraction of the number of times the n_next loop is executed. Second, we already saw in Figure 4-12 that the arithmetic pipeline is the limiting factor in this kernel. The warp scheduler is obviously able to do an excellent job of hiding memory overhead by running it in parallel with computations.

Source	Instructions Executed	Thread Instructions Executed	Thread Execution Efficiency	Memory Type	Memory Access Type	Memory Access Size	L2 Transfer Overhead	L1 Above-Ideal Transactions	L1 Transfer Overhead
else if (iin < d_n_trn_inputs) {	2600000	78987500	89.0	Generic, Global	Load	Size32	0.3	0	1.1
j = 2 * ((icase + istart) * d_n_trn_inputs + iin) ;	2437500	73875000	94.7	Generic, Global	Load	Size32	0.3	0	1.1
r_prev = d_trn_data[j] ;	3737500	113275000	94.7	Generic, Global	Load	Size32, Size64	0.8	135250	1.3
i_prev = d_trn_data[j+1] ;	4087500	123150000	94.2	Generic, Global	Load	Size32, Size64	0.8	135250	1.3
}									
else {									
r_prev = 1.0 ; // Bias	50000	50000	3.1						
i_prev = 0.0 ;	25000	25000	3.1						
}									
ihid = blockIdx.z ;	662500	19750000	93.2						
nhid = d_nhid[0] ; // Neurons in this hidden layer	2437500	74062500	95.0	Generic, Global	Load	Size32, Size64	0.2	-150000	0.7
if (only_hidden) { // Next layer is output layer?	812500	24687500	76.0						
n_next = d_ntarg ;	1950000	59250000	95.0	Generic, Global	Load	Size32	0.3	0	1.1
next_weights = d_wout + ihid * d_ntarg_cols ;	5037500	153062500	95.0	Generic, Global	Load	Size32, Size64	0.2	-150000	0.7
k = d_ntarg_cols / 2 ; // Real and imaginary parts are separated…	9262500	281437500	95.0	Generic, Global	Load	Size32	0.3	0	1.1
}									
else {									
n_next = d_nhid[1] ; // This many neur…	0	0		Generic	Load	Size32, Size64		0	
next_weights = d_whid[1] + ihid * d_nhid_cols[1] ; // Their weights…	0	0		Generic	Load	Size32, Size64		0	
k = d_nhid_cols[1] / 2 ;	0	0		Generic	Load	Size32, Size64		0	
}									
delta_ptr = d_this_delta + icase * 2 * n_next ; // Delta for this…	3900000	118500000	95.0	Generic, Global	Load	Size64	0.1	-150000	0.5
rsum = isum = 0.0 ;	812500	24687500	95.0						
for (j=0 ; j<n_next ; j++) {	384962500	11696937500	79.2						
rsum += delta_ptr[2*j] * next_weights[j] +	3457350000	105050250000	95.0	Generic, Global	Load	Size32, Size64	0.2	-118200000	0.7
delta_ptr[2*j+1] * next_weights[j+k] ;									
isum += -delta_ptr[2*j] * next_weights[j+k] +	3521375000	106995625000	95.0	Generic, Global	Load	Size32, Size64	0.2	-118200000	0.7
delta_ptr[2*j+1] * next_weights[j] ;									
}									
j = icase * nhid + ihid ;	325000	9875000	95.0						
drr = d_drr[0][j] ;	4387500	133312500	95.0	Generic, Global	Load	Size64	0.1	-450000	0.5
dii = d_dii[0][j] ;	4387500	133312500	95.0	Generic, Global	Load	Size64	0.1	-450000	0.5
dri = d_dri[0][j] ;	4387500	133312500	95.0	Generic, Global	Load	Size64	0.1	-450000	0.5
r_delta = rsum * drr + isum * dri ;	487500	14812500	95.0						
i_delta = rsum * dri + isum * dii ;	487500	14812500	95.0						
gptr = d_grad_ptr[0] + icase * d_gradlen ; // Gradient of first hidden…	5525000	167875000	95.0	Generic, Global	Load	Size32, Size64	0.2	-300000	0.6
j = 2 * (ihid * (d_n_trn_inputs + 1) + iin) ;	2437500	74062500	95.0	Generic, Global	Load	Size32	0.3	0	1.1
gptr[j] = r_delta * r_prev + i_delta * i_prev ;	2437500	74062500	95.0	Generic, Global	Store	Size32	2.2	298369	3.0
gptr[j+1] = -r_delta * i_prev + i_delta * r_prev ;	2762500	83937500	95.0	Generic, Global	Store	Size32	2.2	298369	3.0

***Figure 4-13.** Source analysis of routine that computes first hidden gradient*

CHAPTER 5

DEEP Operating Manual

This chapter presents a concise operating manual for DEEP 2.0. The first section lists every menu option along with a short description of its purpose and the page number on which more details can be found if the short description is not sufficient.

Menu Options

This chapter covers the menu options.

File Menu Options

Read a database, page 195

> A text file in standard database format (such as Excel CSV) is read. The first line names the variables, and subsequent lines are the data, one case per line. Space, tab, and comma can be used as delimiters. Subsequent training will produce a predictive model by default, not a classifier.

Read a series (Simple), page 196

> A univariate time series is read, and a set of predictor and target variables are computed based on the values of the series, optionally differenced and/or log transformed. Predictive and classification targets are generated.

Read a series (Trend Path), page 199

> A univariate time series is read, and predictor and target variables are computed based on the path through time of short-term linear trends. Predictive and classification targets are generated.

T. Masters, *Deep Belief Nets in C++ and CUDA C: Volume 2*, https://doi.org/10.1007/978-1-4842-3646-8_5

Read a series (Fourier), page 203

A univariate time series is read, and predictor and target variables are computed based on the Fourier coefficients of data in a moving window. Predictive and classification targets are generated.

Read a series (Morlet), page 207

A univariate time series is read, and a set of predictor and target variables are computed based on Morlet wavelets in a moving window. Predictive and classification targets are generated.

Read XY points, page 211

A set of points in a plane are read and their Fourier transform computed as predictors. Predictive and classification targets are generated.

Read MNIST image, page 213

A standard MNIST-format image file is read. The corresponding MNIST label file must be read after the image file is read. Subsequent training will produce a model that is a classifier by default, not a predictive model.

***Read MNIST image* (Fourier)**, page 213

A standard MNIST-format image file is read, and its two-dimensional Fourier transform is computed to generate predictor variables. The corresponding MNIST label file must be read after the image file is read. Subsequent training will produce a model that is a classifier by default, not a predictive model.

Read MNIST labels, page 214

A standard MNIST-format label file is read. The corresponding MNIST image file must be read before the label file is read.

Write activation file, page 215

A text file containing the activation of a specified neuron for all training set cases is written.

Clear all data, page 215

All training data is erased, but a trained model (if it exists) is retained. The purpose of this command is to allow reading a test dataset and evaluating the performance of a trained model on this new dataset.

Print

The currently selected display window (created under the Display menu) is printed. If no window is selected, Print is disabled.

Exit

The program is terminated.

Test Menu Options

This section covers the Test menu options.

Use CUDA (Toggle Yes/No)

This option is enabled only if a CUDA-capable device is present on the computer. If a check mark appears next to this option, the CUDA device will be used for compute-intensive operations. Click this option to toggle the check mark on and off.

Model Architecture, page 215

The number of unsupervised and supervised layers is specified, as well as the number of neurons in each layer. If the data was read with the *Read a database* command or any of the *Series* commands, the model will be predictive by default, predicting numeric values of the target variable(s). If MNIST data was read, the model will be a classifier by default, employing a *SoftMax* output layer to classify according to the labels in the label file.

Database inputs and targets, page 218

The user specifies one or more predictor variables and one or more target variables. If anything other than a database was read, the predictors and targets are predefined and need not be specified by the user. However, the user can still change them through this menu command if desired. During model training, predictors that are constant for all training cases are omitted from the model.

Advanced options

Options of an advanced nature and that would not normally be changed by the user can be set here. In DEEP 2.0 the only such option is the maximum number of threads allowed for non-CUDA threaded computation. The default should be excellent in all practical applications. It cannot be set to more than 64 because of limitations imposed by the Windows operating system.

RBM training params, page 219

Parameters relevant to unsupervised RBM training can be set.

Supervised training params, page 224

Parameters relevant to training the supervised layers can be set.

Autoencoding training params, page 227

Parameters relevant to autoencoding training can be set.

Train, page 229

The model is trained using the data currently present.

Test, page 232

The trained model is tested with the data currently present.

Cross validate, page 233

The model is evaluated by means of cross validation.

Analyze, page 236

Two basic analyses of the trained model are performed. These are the mean activation of inputs compared to those for the reconstructed data, and the mean activation of the final unsupervised layer. This is valid only if an RBM layer is present.

Display Menu Options

This section covers the Display menu options.

Receptive field, page 237

A plot of the receptive fields (weights of the first/bottom layer) for one or more hidden neurons is displayed. This display may be printed with the File ➤ Print command.

Generative sample, page 238

A plot of one or more generative samples is displayed. This display can be printed with the File ➤ Print command.

Read a Database

A text file in standard database format is read. In particular, standard-format Excel CSV files can be read, as well as databases produced by many common statistical and data analysis programs. The first line must specify the names of the variables in the database. The maximum length of each variable name is 15 characters. The name must start with a letter and may contain only letters, numbers, and the underscore (_) character.

Subsequent lines contain the data, one case per line. Missing data is not allowed.

Spaces, tabs, and commas can be used as delimiters for the first (variable name) and subsequent lines.

Here are the first few lines from a typical data file. Six variables are present, and three cases are shown.

```
RAND0 RAND1 RAND2 RAND3 RAND4 RAND5
-0.82449359  0.25341070  0.30325535 -0.40908301  -0.10667177  0.73517430
-0.47731471 -0.13823473 -0.03947150  0.34984449   0.31303233  0.66533709
 0.12963752 -0.42903802  0.71724504  0.97796118  -0.23133837  0.81885117
```

Read a Series (Simple)

This is the most basic option for reading a time series and automatically generating predictor and target variables. The user specifies a window size, and this window is marched across the time extent of the series. With each placement, as many predictor variables as the window size are generated. Exactly three target variables are generated. The first (used for predictive models) is simply the next value of the predictors, the value one sample past the end of the window. The other two are binary class variables that reflect whether the first target is relatively large or small. The following parameters are specified by the user:

Column: The column of this variable in the database file.

Window: The size of the window placed on the data series. This many predictor variables are generated.

Shift: The number of samples to move the window to generate each case. The default value of one maximizes the number of cases generated. If the value is two, the window will be shifted by two sample points for each placement, thus cutting the number of cases generated roughly in half, and so on.

Nature of variables: raw data: No transformation of any sort is applied to the time series. The predictors and the predicted target are just the values of the time series. This is appropriate when the values of the series itself are predictive and to be predicted, and the standard deviation of the series is at least approximately constant.

Nature of variables: log of raw data: The natural logarithm of each point in the time series is taken. The predictors and the predicted target are these log values. This is appropriate when the values of the series itself are predictive and to be predicted but the data is multiplicative (the average standard deviation during a period in time is proportional to the average value of the series in that period of time). In this situation, taking logs stabilizes the variation across time.

Nature of variables: Difference: Each predictor and the predictive target is computed by taking the difference between adjacent values of the time series. This is appropriate when the changes in the series are predictive and to be predicted, and the variation of these differences is at least roughly constant.

Nature of variables: Difference of log: Each predictor and the predictive target is computed by taking the difference between adjacent values of the natural logarithm of the time series. This is appropriate when the changes in the series are predictive and to be predicted but the values are multiplicative (the average standard deviation of the series during a period in time is proportional to the average value of the series in that period of time). In this situation, taking logs stabilizes the variation across time. The classic use of this variable is for equity prices: the actual price of an equity has no predictive power; rather, it is the changes that are predictive. Moreover, high-priced equities tend to have larger absolute variation than low-priced equities.

Trim tails: Many series have heavy tails; their values are occasionally far from their central tendency. Outliers, whether in predictors or targets, will usually cause severe problems in the training of models when the training algorithm attempts to accommodate the outliers to the detriment of the masses of central values. Setting this to a value greater than zero will cause the specified percent of the largest and the same percent of smallest values to be removed from the database. The input series itself is not affected; trimming applies only to the generated predictors and the predicted target.

Skip header record: If this box is checked, the first record in the series file being read is skipped.

Nature of model: Predict: The model type is set to *predictive* (although this can be overridden by the user in the *Supervised training params* menu). The target variable is set to the predictive target, which is just the next "predictor" past the end of the window. The two class variables are defined by the sign of the predictive target, with *Lead_Pos* being the class if the predictive target is positive, and *Lead_Neg* being the class if the predictive target is zero or negative.

Nature of model: Classify per sign: The model type is set to *classifier* (although this can be overridden by the user in the *Supervised training params* menu). The target variables are set to *Lead_Pos* and *Lead_Neg*. These two class variables are defined by the sign of the predictive target, with *Lead_Pos* being the class if the predictive target is positive, and *Lead_Neg* being the class if the predictive target is zero or negative.

Nature of model: Classify per median: The model type is set to *classifier* (although this can be overridden by the user in the *Supervised training params* menu). The target variables are set to *Lead_Pos* and *Lead_Neg*. These two class variables are defined by the predictive target relative to its median, with *Lead_Pos* being the class if the predictive target exceeds its median, and *Lead_Neg* being the class if the predictive target is less than or equal to its median. If no trained model exists at the time the series is read, this dataset will be used for training, and hence the median of the predictive target will be computed. If a trained model exists at the time the series is read, this new dataset will (presumably) be used for testing, and the median already computed for the presumed training data will be used for defining class membership.

Target multiplier: The target variable is multiplied by this quantity.

As noted, the model type (predictive or classifier) is set according to the user-specified model type, although it can be reset using the *Supervised training params* menu. Similarly, the target variable is set to the predictive value or the two class variables according to the user-specified model type. These, too, can be changed with the *Database inputs and targets* menu. However, it is strongly recommended that the user not tamper with either of these settings. There is rarely any good reason for doing so, and other related default behaviors of the program may be impacted in ways that produce confusing results.

Note that if the model is a classifier and the predictors have almost no predictive power, the trained model will have a strong bias toward classifying cases into whichever class is more prevalent in the training set.

Also note that if the ***Nature of model - Classify per median*** option is chosen, the median of the generated target variable will be computed for defining the class membership of each case if and only if no trained model currently exists. If a trained model already exists, the median will not be recomputed, as the assumption is that the new dataset will serve as an independent test set. In particular, bear in mind that even if a new model is trained with newly read data, the median will *not* be recomputed, because at the time the series was read the program would have been unable to read your mind and know in advance that you will be using the new data to train a new model!

Read a Series (Trend Path)

This is a more advanced option for reading a time series and automatically generating predictor and target variables. The philosophy is described on page 34. The user specifies a window size, and this window is marched across the time extent of the series. With each placement, the linear trend with a fixed lookback is computed for each observation in the window. Each of these generates a predictor variable. Optionally, the difference between these trend values is also computed to create predictor variables.

To be clear, suppose we specify a window size of 5 and a lookback of 20. Suppose we are at time 0. At a minimum, five predictors will be generated. These are the linear trend over the 20-observation time span ending at time 0, that ending at time -1, and those ending at times -2, -3, and -4. Optionally, five more predictors can be generated. The first would be the linear trend ending at time 0 minus that ending at time -1. The second would be that ending at time -1 minus that ending at -2, and so forth. One could consider these changes to be the instantaneous *velocity* of trend change as time passes.

Exactly three target variables are generated. The first (used for predictive models) is based on the observation (the raw input series) one sample past the end of the window. It can be the actual value, the log of the value, or the change from the last observation in the window to the next observation after, or the difference of the logs of these quantities. The other two targets are binary class variables that reflect whether the first target is relatively large or small. In other words, these targets are the same as the targets in the *Simple Series* described in the prior section.

The following parameters are specified by the user:

Column: The column of this variable in the database file.

Window: The size of the window placed on the data series. This many predictor variables are generated if the user chooses to compute values only. Twice this many predictors are generated if the user optionally chooses to include velocity as well.

Shift: The number of samples to move the window to generate each case. The default value of one maximizes the number of cases generated. If the value is two, the window will be shifted by two sample points for each placement, thus cutting the number of cases generated roughly in half, and so on.

Nature of target: raw data: The predicted target is just the next value of the time series. This is appropriate when the values of the series itself are to be predicted and the standard deviation of the series is at least approximately constant. The trends for predictors are based on the raw input series.

Nature of target: log of raw data: The predicted target is the log of the next value of the series. This is appropriate when the values of the series itself are to be predicted but the data is multiplicative (the average standard deviation during a period in time is proportional to the average value of the series in that period of time). In this situation, taking logs stabilizes the variation across time. The trends for predictors are based on the log of the raw input series.

Nature of target: Difference: Each predictive target is computed by taking the difference between the next value of the time series just past the window, and the last value in the window. This is appropriate when the changes in the series are to be predicted and the variation of these differences is at least roughly constant. The trends for predictors are based on the raw input series.

Nature of variables: Difference of log: Each predictive target is computed by taking the difference between the log of the next value of the time series just past the window and the log of the last value in the window. This is appropriate when the changes in the series are to be predicted but the values are multiplicative (the average standard deviation of the series during a period in time is proportional to the average value of the series in that period of time). In this situation, taking logs stabilizes the variation across time. The classic use of this variable is for equity prices: what we really want to predict is the change in price from today to tomorrow. Moreover, high-priced equities tend to have larger variation than low-priced equities. The trends for predictors are based on the log of the raw input series.

Trim tails: Many series have heavy tails; their values are occasionally far from their central tendency. Outliers, whether in predictors or targets, will usually cause severe problems in the training of models when the training algorithm attempts to accommodate the outliers to the detriment of the masses of central values. Setting this to a value greater than zero will cause the specified percent of the largest and the same percent of smallest values *of the target only* to be removed from the database. The input series itself is not affected, nor are the predictors; trimming applies only to the target.

Skip header record: If this box is checked, the first record in the series file being read is skipped.

Nature of model: Predict: The model type is set to *predictive* (although this can be overridden by the user in the *Supervised training params* menu). The target variable is set to the predictive target. The two class variables are defined by the sign of the predictive target, with *Lead_Pos* being the class if the predictive target is positive and *Lead_Neg* being the class if the predictive target is zero or negative.

Nature of model: Classify per sign: The model type is set to *classifier* (although this can be overridden by the user in the *Supervised training params* menu). The target variables are set to *Lead_Pos* and *Lead_Neg*. These two class variables are defined by the sign of the predictive target, with *Lead_Pos* being the class if the predictive target is positive, and *Lead_Neg* being the class if the predictive target is zero or negative.

Nature of model*: *Classify per median: The model type is set to *classifier* (although this can be overridden by the user in the *Supervised training params* menu). The target variables are set to *Lead_Pos* and *Lead_Neg*. These two class variables are defined by the predictive target relative to its median, with *Lead_Pos* being the class if the predictive target exceeds its median, and *Lead_Neg* being the class if the predictive target is less than or equal to its median. If no trained model exists at the time the series is read, this dataset will be used for training, and hence the median of the predictive target will be computed. If a trained model exists at the time the series is read, this new dataset will (presumably) be used for testing, and the median already computed for the presumed training data will be used for defining class membership.

Lookback: The is the number of observations in the source series that will be used for computing linear trend.

Velocity: If this box is checked, then in addition to the *Window* predictors defined as the linear trends, the changes in the trends (the instantaneous velocity) will also be computed as predictors. Thus, there will be twice as many predictors as the window size. It is usually legitimate to treat trend/velocity pairs as real/imaginary pairs as inputs to a complex-domain model. Think of a sine wave and its derivative.

Target multiplier: The target variable is multiplied by this quantity.

As noted, the model type (predictive or classifier) is set according to the user-specified model type, although it can be reset using the *Supervised training params* menu. Similarly, the target variable is set to the predictive value or the two class variables according to the user-specified model type. These, too, can be changed with the *Database inputs and targets* menu. However, it is strongly recommended that the user not tamper with either of these settings. There is rarely any good reason for doing so, and other related default behaviors of the program may be impacted in ways that produce confusing results.

Note that if the model is a classifier and the predictors have almost no predictive power, the trained model will have a strong bias toward classifying cases into whichever class is more prevalent in the training set.

Also note that if the ***Nature of model - Classify per median*** option is chosen, the median of the generated target variable will be computed for defining the class membership of each case if and only if no trained model currently exists. If a trained model already exists, the median will not be recomputed, as the assumption is that the new dataset will serve as an independent test set. In particular, bear in mind that even if a new model is trained with newly read data, the median will *not* be recomputed because at the time the series was read, the program would have been unable to read your mind and know in advance that you will be using the new data to train a new model!

Read a Series (Fourier)

This is a very advanced option for reading a time series and automatically generating predictor and target variables. The philosophy is described on page 41. The user specifies a window size, and this window is marched across the time extent of the series. With each placement, the Fourier coefficients are computed for the data that is in the window. These define a set of predictor variables.

Exactly three target variables are generated. The first (used for predictive models) is based on the observation (the raw input series) one sample past the end of the window. It can be the actual value, the log of the value, or the change from the last observation in the window to the next observation after, or the difference of the logs of these quantities. The other two targets are binary class variables that reflect whether the first target is relatively large or small. In other words, these targets are the same as the targets in the *Simple Series* described in an earlier section.

The following parameters are specified by the user:

Column: The column of this variable in the database file.

Window: The size of the window placed on the data series. Approximately this many predictor variables are generated.

Shift: The number of samples to move the window to generate each case. The default value of one maximizes the number of cases generated. If the value is two, the window will be shifted by two sample points for each placement, thus cutting the number of cases generated roughly in half, and so on.

Nature of target: raw data: The predicted target is just the next value of the time series. This is appropriate when the values of the series itself are to be predicted and the standard deviation of the series is at least approximately constant. The predictors are based on the raw input series.

Nature of target: log of raw data: The predicted target is the log of the next value of the series. This is appropriate when the values of the series itself are to be predicted, but the data is multiplicative (the average standard deviation during a period in time is proportional to the average value of the series in that period of time). In this situation, taking logs stabilizes the variation across time. The predictors are based on the log of the raw input series.

Nature of target: Difference: Each predictive target is computed by taking the difference between the next value of the time series just past the window and the last value in the window. This is appropriate when the changes in the series are to be predicted and the variation of these differences is at least roughly constant. The predictors are based on the raw input series.

Nature of target: Difference of log: Each predictive target is computed by taking the difference between the log of the next value of the time series just past the window and the log of the last value in the window. This is appropriate when the changes in the series are to be predicted but the values are multiplicative (the average standard deviation of the series during a period in time is proportional to the average value of the series in that period of time). In this situation, taking logs stabilizes the variation across time. The classic use of this variable is for equity prices: what we really want to predict is the change in price from today to tomorrow. Moreover, high-priced equities tend to have larger variation than low-priced equities. The predictors are based on the log of the raw input series.

Trim tails: Many series have heavy tails; their values are occasionally far from their central tendency. Outliers, whether in predictors or targets, will usually cause severe problems in the training of models when the training algorithm attempts to accommodate the outliers to the detriment of the masses of central values. Setting this to a value greater than zero will cause the specified percent of the largest and the same percent of smallest values *of the target only* to be removed from the database. The input series itself is not affected, nor are the predictors; trimming applies only to the target.

Skip header record: If this box is checked, the first record in the series file being read is skipped.

Nature of model: Predict: The model type is set to *predictive* (although this can be overridden by the user in the *Supervised training params* menu). The target variable is set to the predictive target. The two class variables are defined by the sign of the predictive target, with *Lead_Pos* being the class if the predictive target is positive, and *Lead_Neg* being the class if the predictive target is zero or negative.

Nature of model: Classify per sign: The model type is set to *classifier* (although this can be overridden by the user in the *Supervised training params* menu). The target variables are set to *Lead_Pos* and *Lead_Neg*. These two class variables are defined by the sign of the predictive target, with *Lead_Pos* being the class if the predictive target is positive, and *Lead_Neg* being the class if the predictive target is zero or negative.

Nature of model: Classify per median: The model type is set to *classifier* (although this can be overridden by the user in the *Supervised training params* menu). The target variables are set to *Lead_Pos* and *Lead_Neg*. These two class variables are defined by the predictive target relative to its median, with *Lead_Pos* being the class if the predictive target exceeds its median, and *Lead_Neg* being the class if the predictive target is less than or equal to its median. If no trained model exists at the time the series is read, this dataset will be used for training, and hence the median of the predictive target will be computed. If a trained model exists at the time the series is read, this new dataset will (presumably) be used for testing, and the median already computed for the presumed training data will be used for defining class membership.

Center: If this box is checked, the data will be centered to have zero mean. This is almost always the best choice. If this is not checked and the data in a window happens to be significantly offset from zero, application of the mandatory Welch data window will introduce spurious low-frequency components. This is discussed in more detail on page 41 and the following pages.

Target multiplier: The target variable is multiplied by this quantity.

As noted, the model type (predictive or classifier) is set according to the user-specified model type, although it can be reset using the *Supervised training params* menu. Similarly, the target variable is set to the predictive value or the two class variables according to the user-specified model type. These, too, can be changed with the *Database inputs and targets* menu. However, it is strongly recommended that the user not tamper with either of these settings. There is rarely any good reason for doing so, and other related default behaviors of the program may be impacted in ways that produce confusing results.

Note that if the model is a classifier and the predictors have almost no predictive power, the trained model will have a strong bias toward classifying cases into whichever class is more prevalent in the training set.

Also note that if the ***Nature of model - Classify per median*** option is chosen, the median of the generated target variable will be computed for defining the class membership of each case if and only if no trained model currently exists. If a trained model already exists, the median will not be recomputed, as the assumption is that the new dataset will serve as an independent test set. In particular, bear in mind that even if a new model is trained with newly read data, the median will *not* be recomputed because at the time the series was read, the program would have been unable to read your mind and know in advance that you will be using the new data to train a new model!

The generated predictors will have names of the form *Real_k* and *Imag_k*, where k is the index of the Fourier coefficient and k ranges from 1 through $n/2$. (If n is odd, $n/2$ means half of n, with the fraction discarded. So, for example, 15/2=7 in this context.) If the data is not centered, one more predictor called *Offset* will be generated. This is

Re(0), the post-windowing data mean. Note that Im(0) is always zero, so no predictor of this name will be generated. Also note that if n is even, Im(n/2) is always zero, so no imaginary predictor will be generated for this quantity.

The model inputs are preset so that they occur in real/imaginary pairs. Thus, *Offset* will not be a preset indicator. Also, the Nyquist pair will be preset only if the window is an odd length. Naturally, the user can manually change these presets if desired.

Read a Series (Morlet)

This is a very advanced option for reading a time series and automatically generating predictor and target variables. The philosophy is described on page 55. The user specifies a window size, and this window is marched across the time extent of the series. With each placement, the real and imaginary Morlet wavelet coefficients are computed for the data that is in the window. These define a set of predictor variables.

Exactly three target variables are generated. The first (used for predictive models) is based on the observation (the input series) one sample past the end of the window. It can be the actual value, the log of the value, the change from the last observation in the window to the next observation after, or the difference of the logs of these quantities. The other two targets are binary class variables that reflect whether the first target is relatively large or small. In other words, these targets are the same as the targets in the *Simple Series* described in an earlier section.

The following parameters are specified by the user:

Column: The column of this variable in the database file.

Window: The size of the window placed on the data series. Twice this many predictor variables are generated (real and imaginary for each position in the window).

Shift: The number of samples to move the window to generate each case. The default value of one maximizes the number of cases generated. If the value is two, the window will be shifted by two sample points for each placement, thus cutting the number of cases roughly in half, and so on.

Period: This is the period over which the wave phenomenon repeats. It must be at least two.

Width: This is the time-domain width of the filter, the number of points on *each side* of the center point. In some contexts, this quantity is called the *half-width*. The total number of points examined to compute a single Morlet wavelet transform value is 2 * *width* + 1. Larger values create a more frequency-selective filter. For the user's protection, DEEP imposes the (not theoretically required) limit that the *width* must be at least the *period*. Setting it to twice the period is a good starting point for experimentation.

Lag: This is the number of samples prior to the current sample at which the filter is centered. This ideally equals the width, and it must not exceed the width. If you are willing to live dangerously, it can be as small as half of the width. Smaller values, even down to zero, are legal but strongly discouraged because of the huge distortion to the filter's frequency response.

Nature of target: raw data: The predicted target is just the next value of the time series. This is appropriate when the values of the series itself are to be predicted, and the standard deviation of the series is at least approximately constant. The predictors are based on the raw input series.

Nature of target: log of raw data: The predicted target is the log of the next value of the series. This is appropriate when the values of the series itself are to be predicted, but the data is multiplicative (the average standard deviation during a period in time is proportional to the average value of the series in that period of time). In this situation, taking logs stabilizes the variation across time. The predictors are based on the log of the raw input series.

Nature of target: Difference: Each predictive target is computed by taking the difference between the next value of the time series just past the window, and the last value in the window. This is appropriate when the changes in the series are to be predicted, and the variation of these differences is at least roughly constant. The predictors are based on the raw input series.

Nature of target*: *Difference of log: Each predictive target is computed by taking the difference between the log of the next value of the time series just past the window and the log of the last value in the window. This is appropriate when the changes in the series are to be predicted but the values are multiplicative (the average standard deviation of the series during a period in time is proportional to the average value of the series in that period of time). In this situation, taking logs stabilizes the variation across time. The classic use of this variable is for equity prices: what we really want to predict is the change in price from today to tomorrow. Moreover, high-priced equities tend to have larger variation than low- priced equities. The predictors are based on the log of the raw input series.

Trim tails: Many series have heavy tails; their values are occasionally far from their central tendency. Outliers, whether in predictors or targets, will usually cause severe problems in the training of models when the training algorithm attempts to accommodate the outliers to the detriment of the masses of central values. Setting this to a value greater than zero will cause the specified percent of the largest and the same percent of smallest values *of the target only* to be removed from the database. The input series itself is not affected, nor are the predictors; trimming applies only to the target.

Skip header record: If this box is checked, the first record in the series file being read is skipped.

Nature of model*: *Predict: The model type is set to *predictive* (although this can be overridden by the user in the *Supervised training params* menu). The target variable is set to the predictive target. The two class variables are defined by the sign of the predictive target, with *Lead_Pos* being the class if the predictive target is positive, and *Lead_Neg* being the class if the predictive target is zero or negative.

Nature of model: Classify per sign: The model type is set to *classifier* (although this can be overridden by the user in the *Supervised training params* menu). The target variables are set to *Lead_Pos* and *Lead_Neg*. These two class variables are defined by the sign of the predictive target, with *Lead_Pos* being the class if the predictive target is positive, and *Lead_Neg* being the class if the predictive target is zero or negative.

Nature of model: Classify per median: The model type is set to *classifier* (although this can be overridden by the user in the *Supervised training params* menu). The target variables are set to *Lead_Pos* and *Lead_Neg*. These two class variables are defined by the predictive target relative to its median, with *Lead_Pos* being the class if the predictive target exceeds its median, and *Lead_Neg* being the class if the predictive target is less than or equal to its median. If no trained model exists at the time the series is read, this dataset will be used for training and hence the median of the predictive target will be computed. If a trained model exists at the time the series is read, this new dataset will (presumably) be used for testing, and the median already computed for the presumed training data will be used for defining class membership.

Target multiplier: The target variable is multiplied by this quantity.

As noted, the model type (predictive or classifier) is set according to the user-specified model type, although it can be reset using the *Supervised training params* menu. Similarly, the target variable is set to the predictive value or the two class variables according to the user-specified model type. These, too, can be changed with the *Database inputs and targets* menu. However, it is strongly recommended that the user not tamper with either of these settings. There is rarely any good reason for doing so, and other related default behaviors of the program may be impacted in ways that produce confusing results.

Note that if the model is a classifier and the predictors have almost no predictive power, the trained model will have a strong bias toward classifying cases into whichever class is more prevalent in the training set.

Also note that if the ***Nature of model - Classify per median*** option is chosen, the median of the generated target variable will be computed for defining the class membership of each case if and only if no trained model currently exists. If a trained model already exists, the median will not be recomputed, as the assumption is that the new dataset will serve as an independent test set. In particular, bear in mind that even if a new model is trained with newly read data, the median will *not* be recomputed because at the time the series was read, the program would have been unable to read your mind and know in advance that you will be using the new data to train a new model!

The generated predictors will have names of the form *Real_k* and *Imag_k*, where *k* is the lag within the window and *k* ranges from 0 (the current point) through one less than the window length.

Read XY Points

This is a very advanced option for reading sets of XY points in a plane and automatically generating predictor and target variables. The philosophy is described on page 69. Each case must contain the same number of points. Each point is represented by a pair of numbers, the X coordinate followed by the Y coordinate. Fourier coefficients with any of several optional normalizations are computed for each case. These define a set of predictor variables. The last item on each line of the file is an integer specifying the class of the case. The class number must range from zero through one less than the number of classes.

One "target" variable is defined for each class. For each case, the target variable for that case's class is set to 1.0 and that for all other classes is set to 0.0.

If the user request using the raw data, the predictors will have names of the form *X_k* and *Y_k* for *k* from 0 through one less than the number of points. Fourier predictors will have names of the form *Real_k* and *Imag_k*, where *k* is the index of the Fourier coefficient and *k* ranges from 0 through one less than the number of points.

The following parameters are specified by the user:

> ***Points***: Each case (line in the file) contains this many points. Each point is a pair of numbers, with the X coordinate coming first, followed by the Y coordinate.

Classes: This is the number of classes. The last number in each line of the file is the class number of the case, ranging from 0 through *Classes*–1. Each line of the file will thus contain 2 * *Points* + 1 numbers.

Type of Data: Raw XY: The (X, Y) points as read from the file are used as default predictors.

Type of Data: Raw Fourier: All Fourier coefficients of the XY points are used as default predictors.

Type of data: Fourier, location normalized: All Fourier coefficients of the XY points except Re[0] and Im[0] are used as default predictors.

Type of data: Fourier, scale/start normalized: All Fourier coefficients of the XY points except Re[0], Im[0], and the real and imaginary parts of the dominant fundamental are used as default predictors. The coefficients that are used are normalized to be location invariant, scale invariant, and starting-point invariant.

Rotation direction: Force CW: The user guarantees that the direction of perimeter tracing defined by the supplied points is clockwise. This option is taken into account only for scale/start normalization. If the user specifies this incorrectly, serious computational errors will result.

Rotation direction: Force CCW: The user guarantees that the direction of perimeter tracing defined by the supplied points is counter-clockwise. This option is taken into account only for scale/start normalization. If the user specifies this incorrectly, serious computational errors will result.

Rotation direction: Auto rotation: The program automatically computes the rotation direction for each case. However, in most applications it is important that this be consistent, the same for every case. If different cases have different rotation directions, chances are your data is invalid, most likely because the perimeter crosses itself. This option is taken into account only for scale/start normalization.

Max frequency: This is the maximum frequency that goes into the default predictor list. All frequencies are put into the database, so the user can manually change this later if desired.

Read MNIST Image

A standard MNIST image file is read. It is assumed that there will be ten labels. The number of rows and columns is read from the file and not assumed by DEEP, although the common file is 28 rows and columns. In DEEP 1.0 the product of the number of rows and columns must not exceed 4096-10=4086. There is no hard-coded limit on the number of images; it is limited only by available memory.

Models in DEEP 2.0 can be either classifiers, in which case the output layer is SoftMax, or predictive, in which case the output layer is linear with no range limiting, and it makes numeric predictions. When MNIST data is read, the classifier form of model is used by default. For all other data, the default is numeric prediction. In both cases, the user can override the default and force the model to be a classifier or predictive.

The MNIST image file must be read before a label file can be read.

Read MNIST Image (Fourier)

A standard MNIST image file is read and its two-dimensional Fourier transform computed to generate predictor variables. It is assumed that there will be ten labels. The number of rows and columns is read from the file and not assumed by DEEP, although the common file is 28 rows and columns. In DEEP 2.0, the product of the number of rows and columns must not exceed 4096-10=4086. Also in DEEP 2.0, the number of rows and columns must be even. There is no hard-coded limit on the number of images; it is limited only by available memory.

Models in DEEP 2.0 can be either classifiers, in which case the output layer is SoftMax, or predictive, in which case the output layer is linear with no range limiting, and it makes numeric predictions. When MNIST data is read, the classifier form of model is used by default. For all other data, the default is numeric prediction. In both cases, the user can override the default and force the model to be a classifier or predictive.

The MNIST image file must be read before a label file can be read.

The names of the transform variables begin with either the letter *R* for a real part or *I* for an imaginary part, followed by the horizontal frequency (never negative) and finally followed by the vertical frequency (also never negative). For example, the variable R_3_7 is the real part of the coefficient for a horizontal frequency of 3 and a vertical frequency of 7. Note the following properties, which are discussed in more detail on page 85:

- The horizontal frequency ranges from zero through the number of columns minus one.
- The vertical frequency ranges from zero through half of the number of rows (the Nyquist frequency). This is in deference to the symmetry depicted in Figure 3-1.
- At a vertical frequency of zero, as well as at the vertical Nyquist frequency, the coefficients at a horizontal frequency of zero and at the horizontal Nyquist frequency are strictly real. Therefore, no actual imaginary parts will be produced as variables for these four coefficients. However, so that "complex" pairs are produced to facilitate complex-domain processing, for these four variables DEEP will generate names starting with *Z* and whose values duplicate the real parts of these four numbers.
- At a vertical frequency of zero, as well as at the vertical Nyquist frequency, the coefficients in the horizontal direction are symmetric (complex conjugates) around the horizontal Nyquist frequency. Therefore, for these two rows, only coefficients through half of the number of columns are generated.

Read MNIST Labels

A standard MNIST label file is read. It is assumed that there are ten possible labels. The label file cannot be read until the image file has been read.

Write Activation File

This option writes a text file containing the activation of a single neuron for all cases, one line per case. The user specifies whether the neuron to be written is in the unsupervised or supervised section, which layer within that section it is in (with 1 being the first layer), and the neuron number within that layer (also with 1 being the first neuron).

An activation file is mainly for diagnostic use, although some users may find it convenient to pass an activation file to other programs.

Clear All Data

Sometimes the user will want to test a trained model on data that the model has not yet seen, often called a test set or out-of-sample (OOS) data. This can be done by reading the training data, training the model, clicking *Clear all data,* reading the test set, and clicking *Test.*

When a trained model exists and data is cleared, subsequently read data must have the same variables in the same order as the data that was used to train the model.

Model Architecture

Figure 5-1 shows the Model architecture dialog.

Model architecture

Model type
- RBM / supervised
- Embedded
- Autoencoder - Real
- Autoencoder - Complex

OK
Cancel

Number of unsupervised layers
This does not include the input layer: 1
Hidden neurons in first unsupervised: 2
Hidden neurons in last unsupervised layer: 2

Number of supervised layers
This includes the output layer: 1
Hidden neurons in first supervised layer: 4
Hidden neurons in last supervised layer: 4

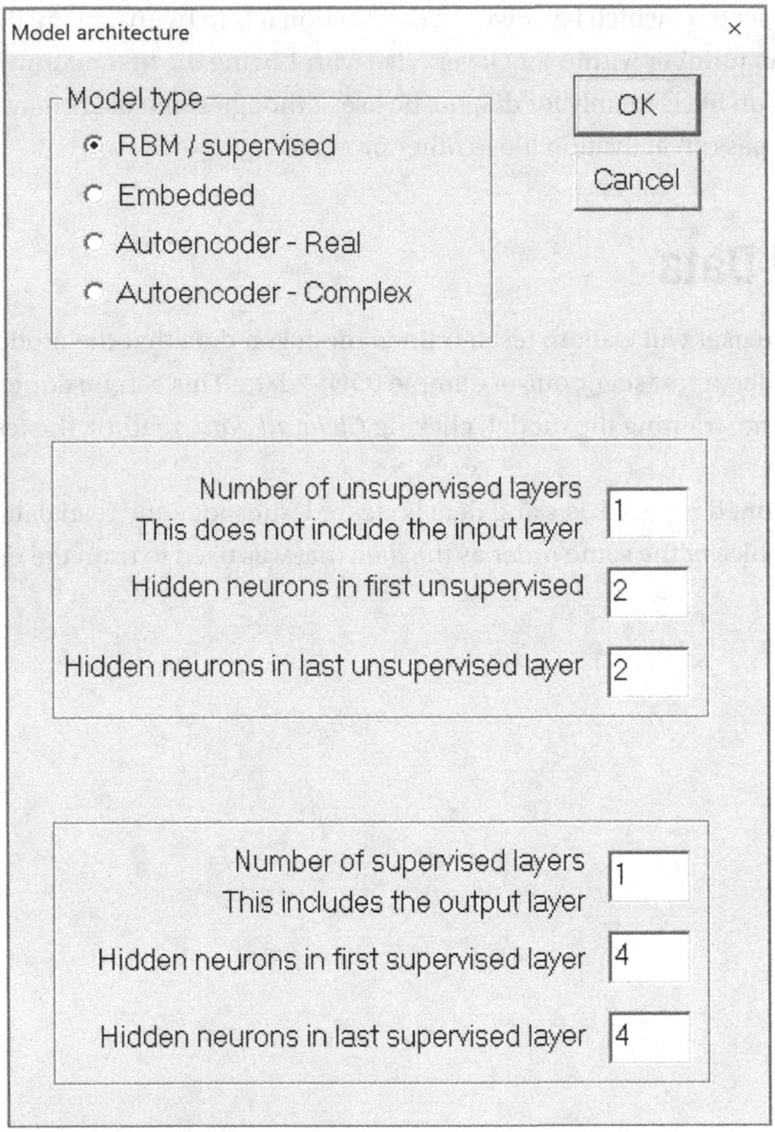

Figure 5-1. *Model architecture dialog*

Several model architectures are available in DEEP 2.0.

- An *RBM / supervised* model consists of zero or more unsupervised layers created by greedy RBM training, followed by one or more supervised layers trained by using the outputs of the final unsupervised layer (or the raw data if there are no unsupervised layers) as inputs and targets as outputs.
- An *Embedded* model consists of a stack of one or more layers (not counting the *input* or *visible* layer). These are greedily trained, as usual for RBMs. The layer just prior to the final (top) layer has class identifier neurons appended, as discussed in Chapter 2.
- An *Autoencoder - Real* model consists of zero or more unsupervised layers created by greedy autoencoder training, followed by one or more supervised layers trained by using the outputs of the final unsupervised layer (or the raw data if there are no unsupervised layers) as inputs and targets as outputs. The entire model operates in the real domain. This family is discussed in depth starting on page 103.
- An *Autoencoder - Complex* model is identical to the previous except that the entire model operates in the complex domain. For the final layer outputs, the imaginary part is ignored; only the real part is used for prediction and classification.

The user defines the architecture by specifying the following quantities:

Number of unsupervised layers: For *Unsupervised/ Supervised* and *Autoencoder* architectures, this may be zero to create a model that is entirely supervised. For *Embedded* architecture, this must be at least one.

Hidden neurons in first unsupervised layer: This refers to the bottom layer, the one closest to the input data.

Hidden neurons in last unsupervised layer: This refers to the topmost unsupervised layer, the one that feeds the supervised section. If there is only one unsupervised layer, this must equal *Hidden neurons in first unsupervised layer*. If there are multiple layers, interior sizes are linearly interpolated.

Number of supervised layers: This is valid only for *Unsupervised / Supervised* and *Autoencoding* architectures. It must be at least one (the output layer), which is the usual case when there are one or more unsupervised layers. But it is legal for an unsupervised section to feed a "traditional" supervised model, one having one or more hidden layers prior to the output layer. It is also possible to use DEEP 2.0 for strictly supervised models.

Hidden neurons in first supervised layer: This is valid only for *Unsupervised / Supervised* and *Autoencoding* architectures. It is relevant only if *Number of supervised layers* is greater than one, in which case it is the number of hidden neurons in the first layer encountered by the unsupervised layer outputs or the raw data if there are no unsupervised layers.

Hidden neurons in last supervised layer: This is valid only for *Unsupervised / Supervised* and *Autoencoding* architectures. It refers to the last hidden layer before the output layer. If *Number of supervised layers* is two (one hidden, plus output), this must equal *Hidden neurons in first supervised layer*. If there are multiple hidden layers (*Number of supervised layers* exceeds two), interior sizes are linearly interpolated.

Database Inputs and Targets

This option is used to specify the variable(s) that will be used as inputs to the model (the *predictors*) and the variable(s) that will be predicted (the *targets*). One or more of each can be selected using standard Windows methods: dragging across a range, holding down *Shift* while clicking the first and last in a range, or holding down *Control* to select individual variables.

If *Read a database* was used to read the training data, then the user must specify the input(s) and target(s). But if the data is anything else, then the inputs and targets are automatically preset. Nonetheless, the user is free to use this menu option to change the preset selection.

All MNIST input variables will follow the naming convention of *P_row_column* to identify the location of each pixel in the input grid, with the naming origin (first row/column) being zero. Thus, the upper-left pixel will be *P_0_0*.

The MNIST target variables will be named *Label_digit* to identify the digit with which each class is associated. Thus, the targets will be named *Label_0* through *Label_9*.

For MNIST data, the model will be a classifier with SoftMax outputs by default. For training data read any other ways, the model will by default be predictive, attempting to predict numeric values for each target. But a supervised training option (described later) allows the user to force the model to be a classifier or predictor. For a forced classifier, the user must specify at least two targets using the *Data inputs and target* menu option, and for each case, the target having maximum value will be assumed to identify the class of the case.

All input selections which compute Fourier transform variables or effective real/imaginary pairs (such as series trend path, which includes velocity) will preselect inputs in such a way that complex-number pairing is obtained. This facilitates immediate use of complex-domain models. Naturally, the user can employ this option to change these presets if desired.

RBM Training Params

Figure 5-2 shows the RBM training parameters.

RBM training parameters ×

Random initialization iterations	50
Number of batches	24
Markov chain length (CD-k) start	1
Markov chain length (CD-k) end	4
Markov chain length (CD-k) rate	0.005000
Learning rate	0.050000
Momentum start (0-1)	0.100000
Momentum end (0-1)	0.900000
Weight penalty (0-1)	0.000100
Sparsity penalty (0-1)	0.001000
Sparsity target (0-1)	0.100000
Increment convergence criterion	0.000010
Max epochs with no improvement	100
Max epochs	10000

Cancel OK

Visible mean field (vs stochastic) ☑
Greedy mean field ☑
Binary split inputs (vs 0-1 rescale) ☐
Fine tune complete model ☐

***Figure 5-2.** RBM training parameters*

This menu option sets the parameters that are relevant to RBM training. All parameters are preset to defaults that should be reasonable for many or most applications. The following parameters may be set:

Random initialization iterations: The number of trial weight sets that are tested to find a good starting point for stochastic gradient descent training. It is definitely worth doing at least a few dozen trials so that subsequent training begins with reconstruction error that is not outrageous. Doing more than several hundred trials is probably overkill.

Number of batches: The training set is divided into this many batches (though the exact number may be adjusted by the program when necessary) for stochastic gradient descent. Concepts vital to this choice are discussed in Volume I. Here are the basic principles:

- Recall from the cited discussion that the trade-off between time-per-batch and batches-for-convergence is unbalanced in the direction of favoring many small batches. But...
- Although Windows threads have fairly small overhead, the overhead of launching a CUDA kernel can be considerable. Thus, one should be inclined to use fewer batches if using CUDA processing.
- The automatic learning rate and momentum adjustment algorithms described in Volume I perform best with relatively large batches. This inspires us to use few batches.
- *This is the most important issue in practice.* Most Windows installations impose an upper limit of two seconds for a CUDA kernel, after which it is given the boot. Kernel time is almost linearly related to batch size, so if your screen blacks out and recovers with a message that the driver was reset, increase the number of batches. CUDA.LOG lists kernel times and hence can be used to see how close to criticality you are.

Markov chain length (CD-k) start: When stochastic gradient descent begins, this is the number of iterations taken by executing the Markov chain in the contrastive divergence algorithm.

The gradient estimate's accuracy is improved by taking more iterations, with the result that convergence requires fewer epochs. But these samples are very expensive to obtain. Early in training we do not need accurate gradient estimates; a rough approximation is sufficient. This parameter should almost always be left at its default value of 1.

Markov chain length (CD-k) end: The number of iterations taken as learning progresses. As convergence nears, it is worthwhile expending computation time to obtain more accurate gradient estimates. The default value of four is good in nearly all applications. In case the user wants to obtain true maximum likelihood parameter estimates (usually pointless in practice), this parameter can be set to a very large value.

Markov chain length (CD-k) rate: The rate at which the chain length increases from the starting value to the end value. Standard exponential smoothing is employed, with the ending chain length being the "new value" of the series.

Learning rate: The initial learning rate. This should be small, probably smaller than the value the user is accustomed to for other programs. This is because the automatic adjustment algorithm described in Volume I will rapidly move it to an optimal value.

Momentum start: The initial momentum. This should be small, probably smaller than the value the user is accustomed to for other programs. As with the learning rate, the automatic adjustment algorithm described in Volume I will rapidly move it to an optimal value.

Momentum end: As training progresses, the momentum will progress toward this value unless the adjustment algorithm swats it down because of instability in the gradient descent algorithm. Values greater than the default are dangerous, and even the default is pretty high.

Weight penalty: The degree to which large weights are penalized. This must be small to allow weights to approach their optimal values. But it should not be zero. If no weight penalty is applied, in unusual but annoying pathological situations one or more weights can blow up to enormous values.

Sparsity penalty: The degree to which hidden neuron activation rates are encouraged to approach the *sparsity target* specified as the next parameter. This is not a critical parameter, and it can safely be set to zero if desired. However, in most cases it is good to gently nudge weights toward values that result in smallish hidden neuron activation rates, such as 0.1 or so. Among other things, this makes the weights more interpretable, as one can then study which patterns are associated with activation of certain hidden neurons. If all hidden neurons are activated about half the time, such interpretation is more difficult than if activation is rarer.

Sparsity target: The value toward which hidden neuron activation rates are nudged by the *sparsity penalty*. This is typically around 0.1 or so. This parameter is ignored if the *sparsity penalty* is zero.

Increment convergence criterion: This is the secondary convergence criterion, as described in Volume I. If the ratio of the magnitude of the largest weight adjustment in an epoch to the magnitude of the largest weight drops below this threshold, convergence is decreed to be complete. This should be very small to avoid early exits from the training algorithm.

Max epochs with no improvement: This is the primary convergence criterion. The ratio of the magnitude of the largest weight adjustment in an epoch to the magnitude of the largest weight is a good (though not perfect) measure of how close we are to a local minimum of the negative log likelihood criterion being minimized. If the specified number of epochs pass without this ratio beating its minimum so far, convergence is said to have been achieved.

Max epochs: This is a backstop, insurance against endless iteration. It should never be used as an actual convergence criterion, as it is a brute-force rule, with no intelligence about actual convergence. Make it large, and trust that except in very rare pathological situations, one of the main convergence criteria will handle the situation well.

Visible mean field (vs stochastic): If this box is checked, the reconstruction of the visible layer will use the mean field approximation. If not checked, the reconstruction will sample. It is likely that using the mean field approximation is best, although this is not universally agreed upon. In practice, the difference seems slight.

Greedy mean field: If this box is checked, propagation of input data through early layers for greedy training strictly uses mean field approximations. If not checked, sampling is done for the inputs to the layer being trained (except the first layer, which is never sampled). This topic is discussed in detail in Volume I.

Binary splits: If this box is checked, the raw input data will be quantized to strictly binary data by setting variables above their mean to one and those equal or below the mean to zero. If not checked, the raw input data will be linearly scaled to a range of 0-1.

Fine tune complete model: If this box is checked, after the entire deep belief net is constructed (all RBMs greedily trained, then all subsequent layers trained with supervision), supervised training will be used to tweak the entire model, including the RBM layers. This will always improve in- sample performance and often improve out-of-sample performance. But display of reconstruction samples becomes pointless garbage.

Supervised Training Params

Figure 5-3 shows the supervised training parameters.

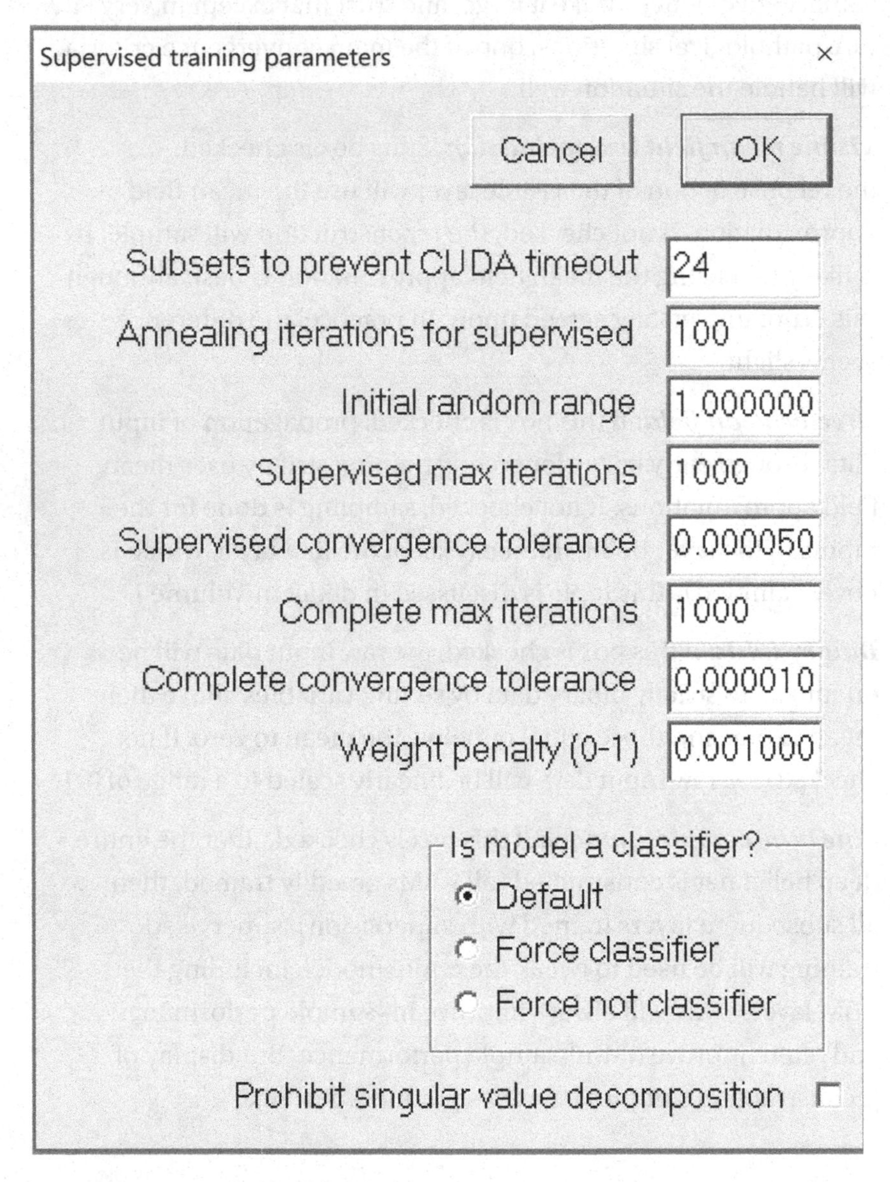

***Figure 5-3.** Supervised training parameters*

This menu option sets the parameters that are relevant to supervised training of the layer(s) following the RBM layer(s), as well as the optional fine-tuning of the complete deep belief net. All parameters are preset to defaults that should be reasonable for many or most applications. The following parameters may be set:

> ***Subsets to prevent CUDA timeout***: This has no effect whatsoever on the model produced. It affects only the degree to which computations are split up; the results of the computations remain the same. *This is different from batches in RBM training.* RBM batch division does impact the model and the nature of convergence because the weights are updated for each batch. In supervised training, all batches are pooled, with one weight update per epoch (pass through the entire training set). To reduce kernel-launch overhead, the number of subsets should be set as low as possible. But keep an eye on the CUDA time summary in CUDA.LOG and be prepared to use more subsets if any time-per-kernel approaches the two-second Windows limit.
>
> ***Annealing iterations for supervised***: The number of simulated annealing passes used to find a good weight set from which to begin training. This topic is discussed in detail in Volume I. This is usually a fairly cheap operation with good returns for the first few hundred passes. More than a few thousand iterations is probably overkill due to rapidly diminishing returns.
>
> ***Initial random range***: The average range of weight perturbation for simulated annealing. The program will periodically raise and lower the user-specified figure to make this parameter less critical. For this reason, the progress plot of error will have clearly visible periodic variation. This is normal operation. The exact algorithms that govern simulated annealing perturbation are shown in Volume I.
>
> ***Supervised max iterations***: After RBM training is complete, the supervised layer(s) following the RBM layer(s) are trained. This parameter limits the number of epochs in order to prevent wildly excessive run times. It should be set to a very large value and used as insurance only, not as the usual convergence determiner.

Supervised convergence tolerance: This is the primary method for determining convergence of training the supervised layer(s). Training is stopped when the relative change in the error from one epoch to the next falls below this level. Because the supervised training algorithm used in DEEP is deterministic, this can safely be set to a very small value, although doing so is usually without merit because most improvement happens early in training.

Complete max iterations: This is identical to *Supervised max iterations* except that it applies to the optional fine-tuning of the complete (unsupervised RBMs plus subsequent supervised layers) deep belief net.

Complete convergence tolerance: This is identical to *Supervised convergence tolerance* except that it applies to the optional fine-tuning of the complete (unsupervised RBMs plus subsequent supervised layers) deep belief net.

Weight penalty: This penalty discourages large weights during supervised training. It should nearly always be set to a very small value, small enough that it does not have an overly strong impact on learning "best" weights, yet large enough that it prevents the large weights that can happen in some unusual pathological situations that are especially likely when the inputs to the supervised section are strongly correlated. This topic is discussed in detail in Volume I.

Is model a classifier: By default, MNIST data produces a classifier model, and all other data produces a predictive model. This option allows the user to override the default. If database data is read and the user forces the model to be a classifier, at least two targets must be selected, and for each case the target having greatest value is assumed to be the correct case.

Prohibit singular value decomposition: The section beginning on page 115 discusses how the extremely efficient singular value decomposition (SVD) algorithm can be used to explicitly compute optimal output weights for a predictive model and discover excellent starting weights for iterative training of classifiers. But for gigantic problems and in some very rare pathological situations, SVD can fail, even (very rarely) producing not-a-number results. For this reason, SVD is disabled if there are more than 400 inputs to the output layer. Moreover, the user may choose to disable SVD. Because SVD is such an enormous help in achieving rapid and high-quality convergence, it should always be allowed if at all possible.

Autoencoding Training Params

Figure 5-4 shows the autoencoding training parameters.

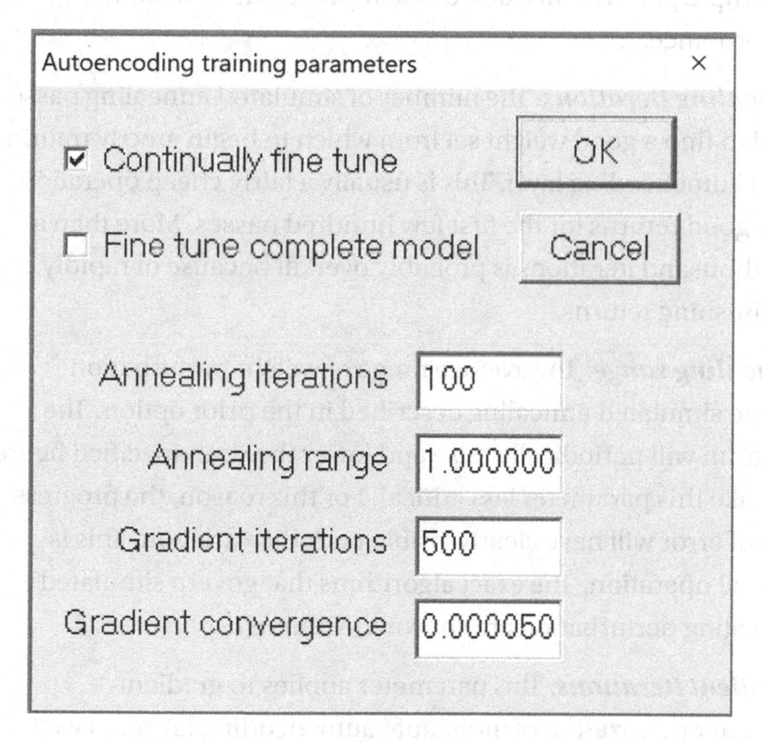

Figure 5-4. *Supervised training parameters*

Continually fine tune: If this box is not checked, each unsupervised autoencoding layer will be trained individually and independently. If this box is checked, after two layers are trained (each separately), these two layers will be pooled, and the two-layer network trained to autoencode the inputs. If a third layer is then added (by independent training), it will then be added to the pool and this three-layer network trained as an input autoencoder, and so on. This algorithm is described in detail on page 108.

Fine tune complete model: If this box is checked, after the entire deep belief net is constructed (all autoencoding layers greedily trained, then all subsequent layers trained with supervision), supervised training will be used to tweak the entire model, including the unsupervised layers. This will always improve in-sample performance and often improve out-of-sample performance.

Annealing iterations: The number of simulated annealing passes used to find a good weight set from which to begin greedy training of an autoencoding layer. This is usually a fairly cheap operation with good returns for the first few hundred passes. More than a few thousand iterations is probably overkill because of rapidly diminishing returns.

Annealing range: The average range of weight perturbation for the simulated annealing described in the prior option. The program will periodically raise and lower the user-specified figure to make this parameter less critical. For this reason, the progress plot of error will have clearly visible periodic variation. This is normal operation. The exact algorithms that govern simulated annealing perturbation are shown in Volume I.

Gradient iterations: This parameter applies to gradient-descent optimization of individual autoencoding layers as well as continuous fine tuning of autoencoding layers. It limits the number of epochs in order to prevent wildly excessive run times. It should be set to a very large value and used as insurance only, not as the usual convergence determiner.

Gradient convergence: This parameter applies to gradient-descent optimization of individual autoencoding layers as well as continuous fine tuning of autoencoding layers. This is the primary method for determining convergence of training the autoencoding layer(s). Training is stopped when the relative change in the error from one epoch to the next falls below this level. Because the training algorithm used in DEEP is deterministic, this can safely be set to a very small value, although doing so is usually without merit because most improvement happens early in training.

Train

The *Train* selection trains the entire deep belief net. If this is an RBM/Supervised model or an autoencoding model, then first all RBM or autoencoding layers are trained with unsupervised greedy training. Then, all subsequent layers (typically just one, the output) are trained using supervision. Finally and optionally, the entire deep belief net is fine-tuned with supervision. The steps for complete training are shown at the left side of the screen. Those that will not be used in the current configuration are grayed out. A marker arrow identifies the step currently executing, and particularly slow operations indicate the percent completion.

The first step in RBM training is finding initial weights by randomly generating weight sets and finding the one with minimum reconstruction error. In Figure 5-5 we see this operation in progress. The top line on the left side says that we are training RBM layer 1. The initial weight operation is 55 percent complete. The graph is the RMS reconstruction error, with the light blue line showing the individual tries and the heavy black line showing the best so far.

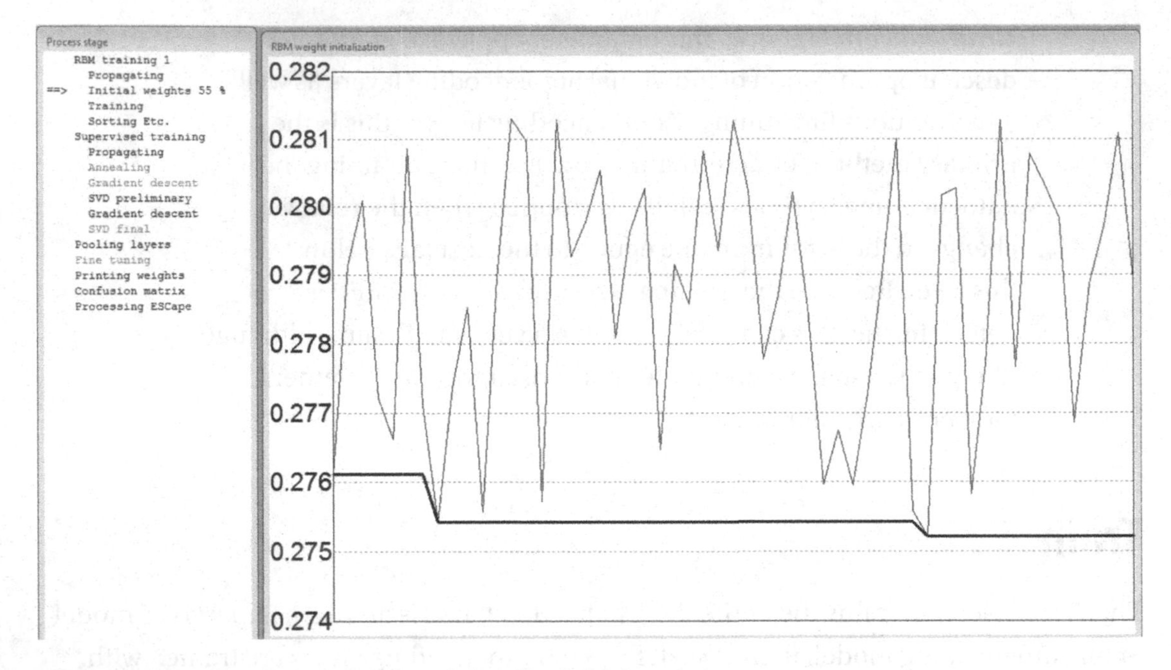

Figure 5-5. *Finding initial weights for RBM training*

After initial weight selection is complete, the program trains the RBM using stochastic gradient descent. The screen will resemble Figure 5-6.

At the left side we see that we are in the *Training* operation and we are 1 percent done. This percentage is relative to the *Max epochs* parameter, which, as stated earlier, should always be set overly large and used only as a backstop. Hence, this percentage will nearly always be very pessimistic relative to actual training progress.

The largest window plots three values, whose current, minimum, and maximum values are written in the top center of the plot. The *reconstruction error* is in red, and it typically drops off fast and then levels out. The *increment ratio* (maximum increment divided by maximum weight) typically decreases fairly linearly before hitting a sharp knee and flattening, with a few subsequent small bounces. The RMS gradient often displays peculiar behavior, with very gradual decrease punctuated by sharp jumps up as blocks of weights suddenly go from near zero to larger, more useful values.

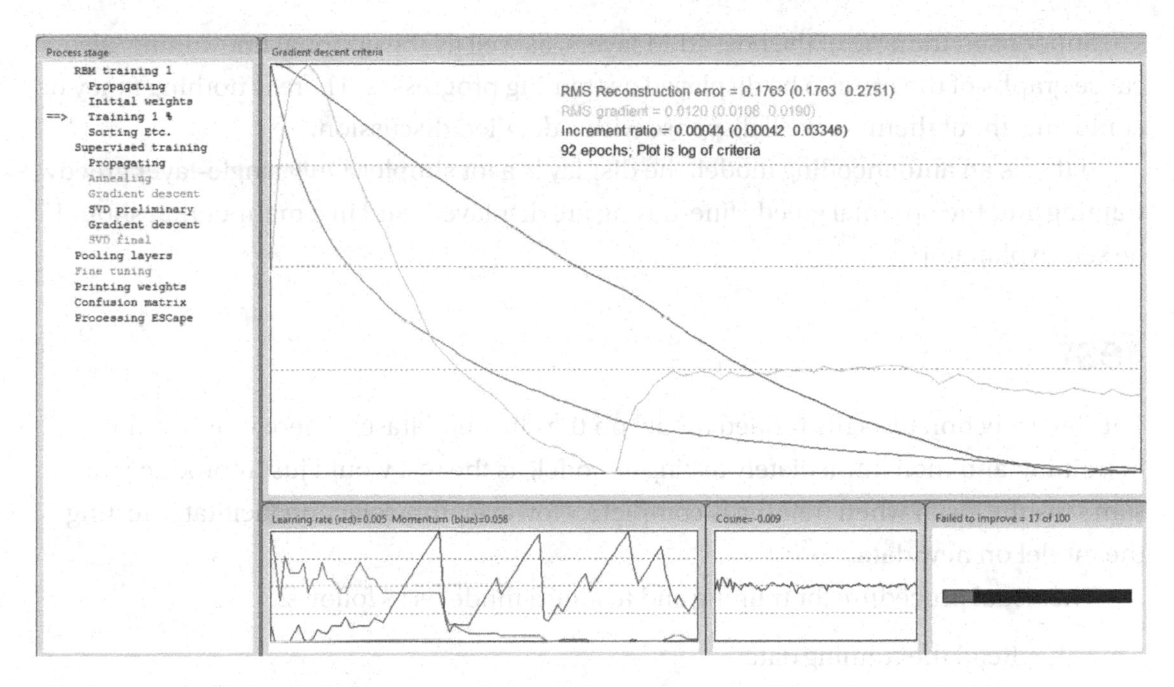

Figure 5-6. *RBM training*

Be aware that these plots are the logs of the values, not the actual values. Also, each plot is scaled so that the entire historical range of the parameter exactly covers the vertical extent of the plot. The net effect is that as training progresses and values become small, tiny changes in the actual values are magnified to large changes in the plot. This magnification is useful in that it shows in great detail exactly what is happening. Unfortunately, it can be deceptive, making the user think that violent gyrations are occurring when, in fact, the changes in the actual values are miniscule.

The lower-left graph shows the dynamically adjusted learning rate and momentum, also scaled so that the historical values exactly fill the vertical extent of the plot. Typically, the learning rate will show a net decrease, dropping to a very small value after several dozen iterations during which it bounces. The momentum only rarely stabilizes, climbing steadily until it becomes excessive and causes an overshoot that results in backtracking, at which point the adjustment algorithm slaps it back down for a while.

The bottom-center graph shows the cosine of the angle between successive gradients, scaled to a fixed range of minus one to one. It should always be near the center, indicating that the weight increments are neither undershooting nor overshooting.

The lower-right bar graph shows the number of contiguous failures of the increment ratio to decrease, relative to the user-specified limit. When the red interior reaches the right side of the bar's outline, training will terminate. This is the primary convergence criterion.

Supervised training of the post-RBM layers, as well as the optional fine-tuning, also cause graphs of the error to be displayed as training progresses. There is nothing fancy or confusing about them, so we'll dispense with a detailed discussion.

If this is an autoencoding model, the display is a lot simpler. Only single-layer greedy training and the optional greedy fine-tuning are displayed, and in a manner that should be self-explanatory.

Test

The *Test* selection tests the trained model on the current dataset. There is little point in training and then immediately testing a model, as the test would just reproduce the same results given when training is complete. However, this selection facilitates testing the model on new data.

The usual procedure for training and testing a model is as follows:

1. Read the training data.
2. Define the architecture.
3. Select the predictor and target variables if they are not preset. (Only Read a Database does not preset them.)
4. Set training parameters if something other than the default is desired.
5. Train.
6. Clear all data.
7. Read the test data.
8. Test.

The test dataset must contain the same variables in the same order as the training dataset. The user must not change the architecture or the predictor/target variables.

Note The *Test* option does not use CUDA processing. If the model was trained with CUDA enhancement, it is possible that the slightly different floating-point computations with and without CUDA may result in slightly different test results. Any differences should be small.

Cross Validate

Figure 5-7 shows the cross-validation settings.

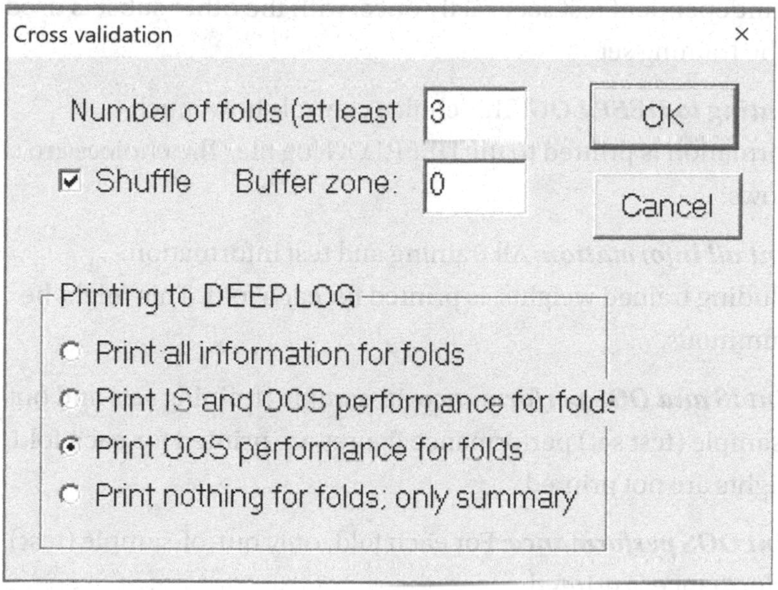

Cross validation ×

Number of folds (at least 3 OK

☑ Shuffle Buffer zone: 0 Cancel

Printing to DEEP.LOG

○ Print all information for folds

○ Print IS and OOS performance for folds

◉ Print OOS performance for folds

○ Print nothing for folds; only summary

Figure 5-7. *Cross-validation*

The model's capability is evaluated by cross validation. In this process, the dataset is divided into a user-specified number of subsets, as equal in size as possible. One at a time, each of these subsets is used as a test set after the model is trained by combining the remaining subsets into a single training set. After each subset has been processed this way, which is called a *fold*, the test set results for all folds are pooled into a single grand performance measure. The beauty of this technique is that we do not have to sacrifice data by designating part of the dataset as a one-time-only training set and the other part as a one-time-only test set. Each case in the dataset serves as a test case exactly once, which is an extremely efficient use of the data! Moreover, each time we train the model we are using the large majority of the training cases, which encourages stability. The fact that the pooled test results are nearly unbiased is almost miraculous. Of course, the price paid is that we must retrain the model for each fold, which can be very expensive in some applications. There's no free lunch.

The user must specify several options, as follows:

Number of folds: The dataset will be divided into this many subsets, as equal in size as possible. Each subset will be used as the independent test set exactly once, with the other subsets used as the training set.

Printing to DEEP.LOG: This choice controls how much information is printed to the DEEP.LOG log file. The choices are as follows:

Print all information: All training and test information, including trained weights, is printed for each fold. This might be voluminous.

Print IS and OOS performance: In-sample (training set) and out-of-sample (test set) performance figures are printed for each fold. Weights are not printed.

Print OOS performance: For each fold, only out-of-sample (test) performance is printed.

Print nothing for folds: No information is printed for individual folds. Only the pooled summary performance report appears.

Shuffle: If this box is checked, the dataset is shuffled before folds are assigned. This is mandatory if the data is "grouped" in any way, perhaps by inherent serial correlation or perhaps by experimental design. Suppose, for example, the design incorporates three experimental conditions, each grouped together, and we employ three folds of cross validation. If the data is not shuffled, then for each fold we will be testing data from a condition that did not appear in the training set for that fold!

Buffer zone: This is required if one or more predictors have serial correlation *and* one or more targets have serial correlation. In this situation, the specified number of cases are temporarily removed from the training set on the upper and lower fold boundaries. The test set is left unaltered. This is illustrated in Figure 5-8. In this figure, there are five folds, delineated with tick marks, and the fold depicted happens to be the one with the test set in the center. Note how part of the training set on each side of the boundary is removed. To determine the correct buffer zone size, find the maximum distance of serial correlation among the predictors and the maximum among the targets. Take the minimum of these two numbers to get the optimal buffer zone.

For example, let the predictors have a maximum correlation distance of 5 cases, and the target 3. The optimal buffer would be Min(5,3)=3. Consider the left boundary. Define time 0 as the first case in the test set. Then consider the following discussion.

Suppose we did not use a guard buffer zone. The last training case at the left boundary would be at time -1. Its targets could be correlated up to the case at time 2. The first test case could have predictors correlated back to time -5, meaning that its predictors could be similar to several cases in the training set. This alone is of no consequence. But recall that the last training case's targets could be similar to those up though the case at time 2. So this first test case can *also* have targets similar to those in the training set. Having an "independent" test case substantially artificially represented in the training set introduces bias.

It's important to understand the problem here. There is nothing wrong with having serial correlation in *either* the predictors alone (a very common situation) *or* the targets alone (not so common). The problem happens when *both* have serial correlation. In this situation, the training cases near the boundary can be substantially similar to the test cases near the same boundary. The test set, which is supposed to be independent of the training set, is actually not totally independent because serial correlation has destroyed independence near the boundaries.

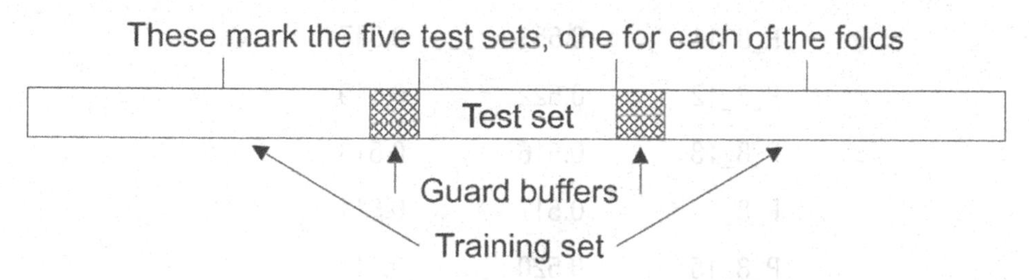

Figure 5-8. *Cross validation with guard buffers*

Now suppose we remove the three guard zone training cases. The last training set case at this boundary is at time -4. Its targets could be correlated with those of cases through time -1. But the target correlation has vanished by the time we get to the first test case. We're safe.

A similar effect happens at the right boundary. Let the first training case in the upper section be at time 0. Then the last test case below it is at time -1, and its targets could be correlated with the targets through the case at time 2. But after removal of the guard zone, the first used training case will be at time 3.

We can reverse the correlation pattern, with the correlation distance for predictors 3 cases and that for targets 5. The guard zone is still 3 cases. At the left boundary, let the first test case be at time 0. Its predictors can be correlated with those of the cases back to time -3. The last "original" training case is at time -1, but after removing the guard zone the last actually used training case is at time -4. I leave it as an exercise for you to check the right boundary.

Analyze

This selection computes and prints to the DEEP.LOG file two tables of information for RBM models (only!). The first is a comparison, for each input variable, of the probability of its being activated in the training set versus the probability of its being activated in the reconstructed input layer. Here is a short segment illustrating this table:

Variable	Visible	Reconstructed
P_8_10	0.616	0.617
P_8_11	0.551	0.547
P_8_12	0.522	0.519
P_8_13	0.516	0.513
P_8_14	0.517	0.511
P_8_15	0.520	0.514
P_8_16	0.517	0.513
P_8_17	0.514	0.510
P_8_18	0.539	0.536

(Continued)

Variable	Visible	Reconstructed
P_8_19	0.606	0.603
P_8_20	0.706	0.706
P_8_21	0.806	0.810
P_8_22	0.887	0.891
P_8_23	0.942	0.943

The other analysis output is the probability (across the training set) of each final (topmost) layer hidden neuron being activated. Here is an example of this table:

Hidden	Activation
1	0.837
2	0.449
3	0.723
4	0.596
5	0.578
6	0.501
7	0.501
8	0.418

Receptive Field

The *receptive field* of a hidden neuron in an RBM is (loosely) defined as the pattern of weights connecting the input layer to the hidden neuron. If the input happens to be an image, such as is the case with MNIST data, then it is possible to display these weights in the same dimensions as the input image. Figure 5-9 shows the receptive fields of a dozen neurons trained with MNIST data. Large positive weights are white, large negative weights are black, and intermediate values are shades of gray. A color display is also an option, with positive weights colored cyan and negative weights colored red, and brightness corresponding to magnitude. The gray areas around the perimeter are pixels that are constant for all cases and hence omitted from the model.

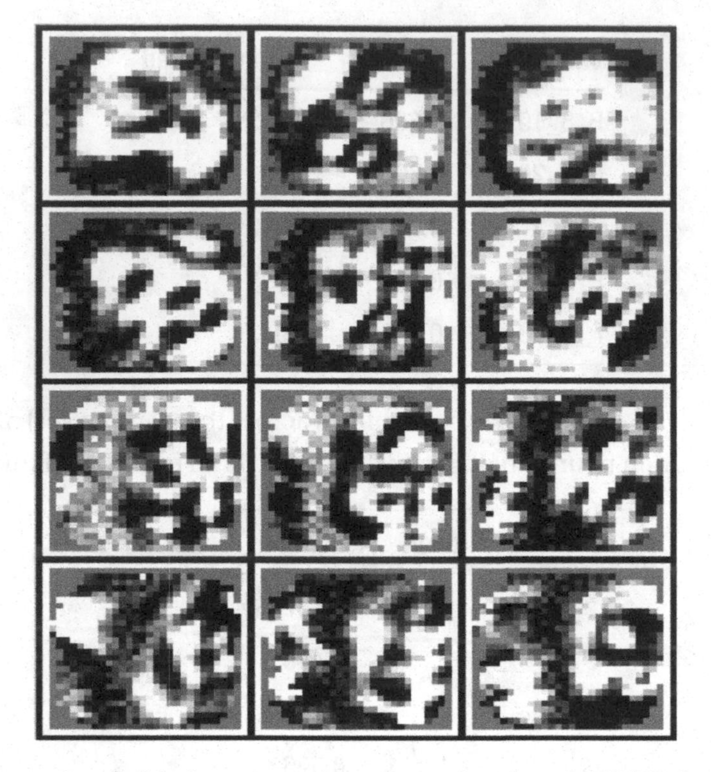

Figure 5-9. *Receptive fields for some neurons trained on MNIST data*

Generative Sample

We saw in Volume I and on page 97 of this book that a trained RBM or set of greedily trained layers can be made to spit out random samples from the distribution that it has learned. Examination of such random samples can be interesting because they show examples of the primitive patterns that the model has learned.

This option is valid only for MNIST images and *simple series* data. The user must specify the number of rows and columns of samples to display. Each of the *nrows*ncolumns* images is a separate sample.

As discussed in Volume I, there are two ways to begin the Markov chain whose final value will be the computed sample. One can begin with a member of the training set. To do this, set the *First case* field to a positive number, the sequential number of the training case that will be used for the first sample. Subsequent samples will start from subsequent training cases. The degree to which the final reconstruction resembles the starting pattern is an indication of the quality of training and the degree to which efficient mixing is taking place in the Markov chain.

Figure 5-10 shows the first 12 cases from the MNIST test set of ten thousand cases. Figure 5-11 shows generative samples obtained from these cases using 10,000 iterations. What makes this interesting is that this was derived from a single RBM layer having just 15 hidden neurons! The degree to which this tiny model has encapsulated training set patterns is astounding.

Alternatively, one can set the topmost hidden neuron layer to random values, thus divorcing the computed samples from training data. This lets us see the actual primitive patterns which the model is recognizing. Figure 5-12 shows 108 random samples obtained from an RBM having 100 hidden neurons, using 50,000 iterations. This is not an *embedded* model, which allow class-conditional sample generation. So rather than often seeing *digits*, we are more likely to see the *components* of the digit images that the model has learned.

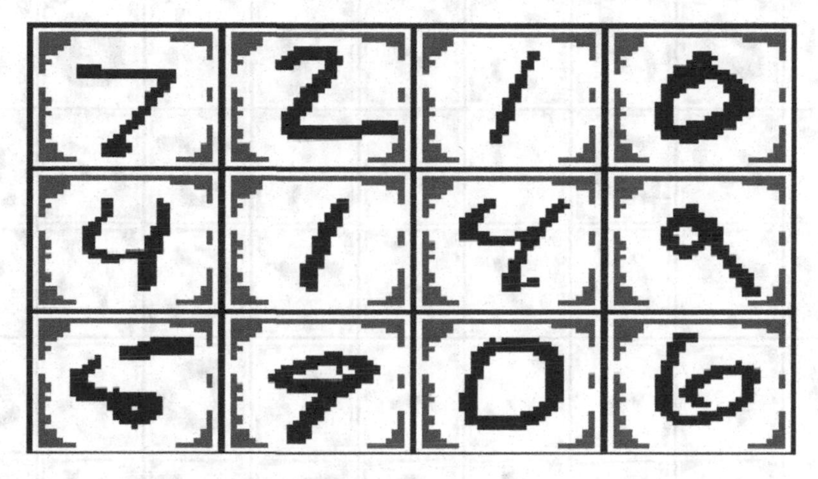

Figure 5-10. *First 12 cases of MNIST test set*

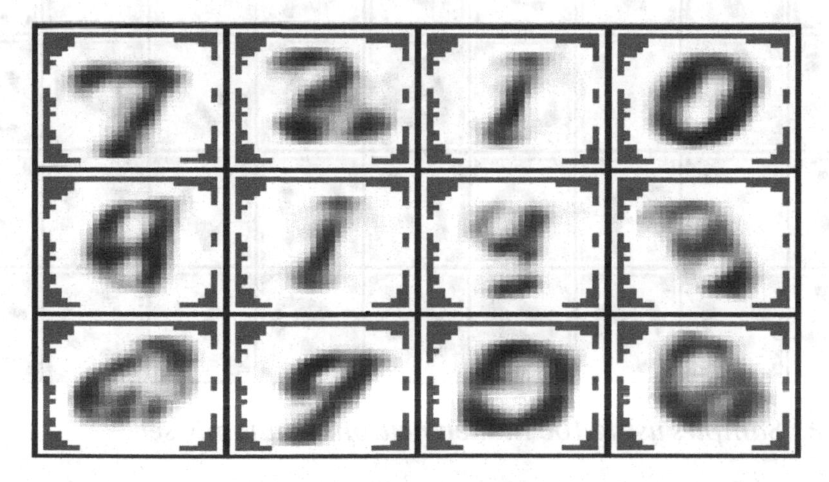

Figure 5-11. *Generative samples after 10,000 iterations*

Figure 5-12. *Samples using 100 hidden neurons randomly set*

Samples from an Embedded Model

We saw on page 1 that by embedding class labels in the visible layer corresponding to the top-level RBM, we can perform generative sampling from each class separately. When an embedded model has been trained, two additional selections may be made by the user.

The first choice is the check box *Clamp class label.* This would almost always be left at its default of checked, as clamping a class label to force class-conditional sampling is usually the point of using an embedded model (at least for me!). Unchecking this box lets generation run free, without regard to class labels, and thus provides samples from the entire population.

The other choice is *Class for Clamped Random.* This is relevant only if the user specified zero for the *First class label.* With zero, the Markov chain is initialized to random values, so the *Class for Clamped Random* specification is needed to tell the program which class to sample. If the *First class label* is positive, then the class of each training case used for initialization will be clamped.

Figure 5-13 shows some generated samples from the MNIST 0 class, and Figure 5-14 shows samples from the 1 class.

If the user specified that binary thresholding be performed for RBM training (*Binary splits* on page 222), then generated samples will also be quantized to binary. Figures 5-15 and 5-16 show binary generated samples from the MNIST 0 and 1 classes, respectively. Note the significant amount of exact or near duplication, indicating that the RBM has learned some "favorite" patterns and converges strongly to them.

If the data input is a simple series (page 22), then the user must specify the *Row resolution.* This is the vertical resolution for the display, the number of rows over which the signal is quantified. The default of 20 is reasonable for most situations, although it can be argued that setting the row resolution approximately equal to the window length is also a good choice. An example of generative samples of a simple series can be seen in Figure 2-3 on page 33.

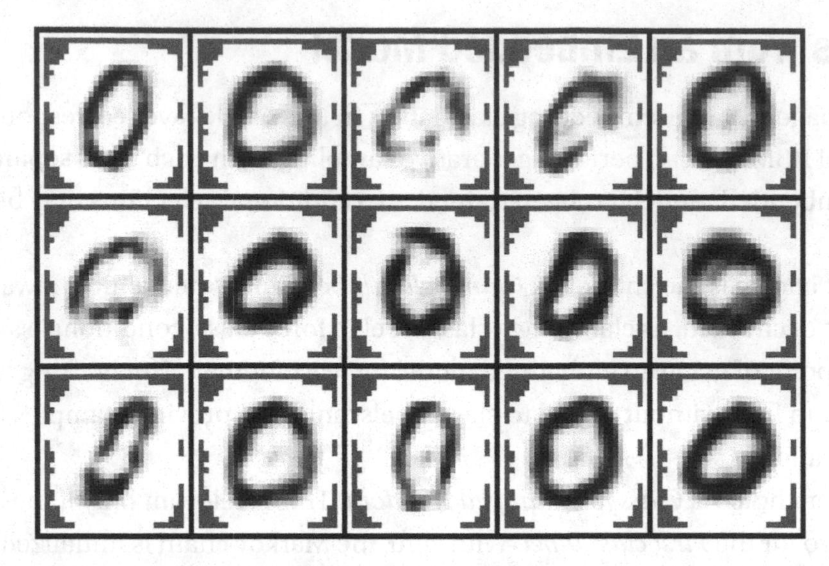

Figure 5-13. *Generative samples from MNIST 0 class*

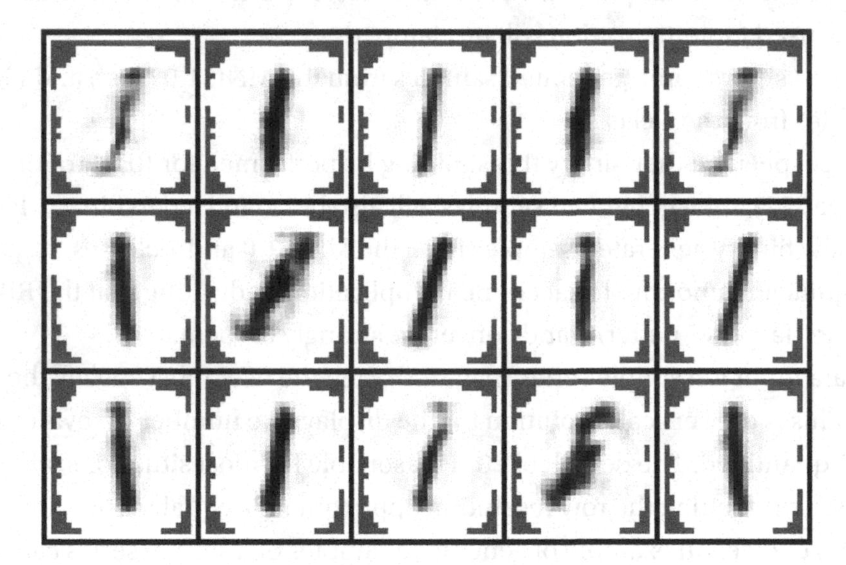

Figure 5-14. *Generative samples from MNIST 1 class*

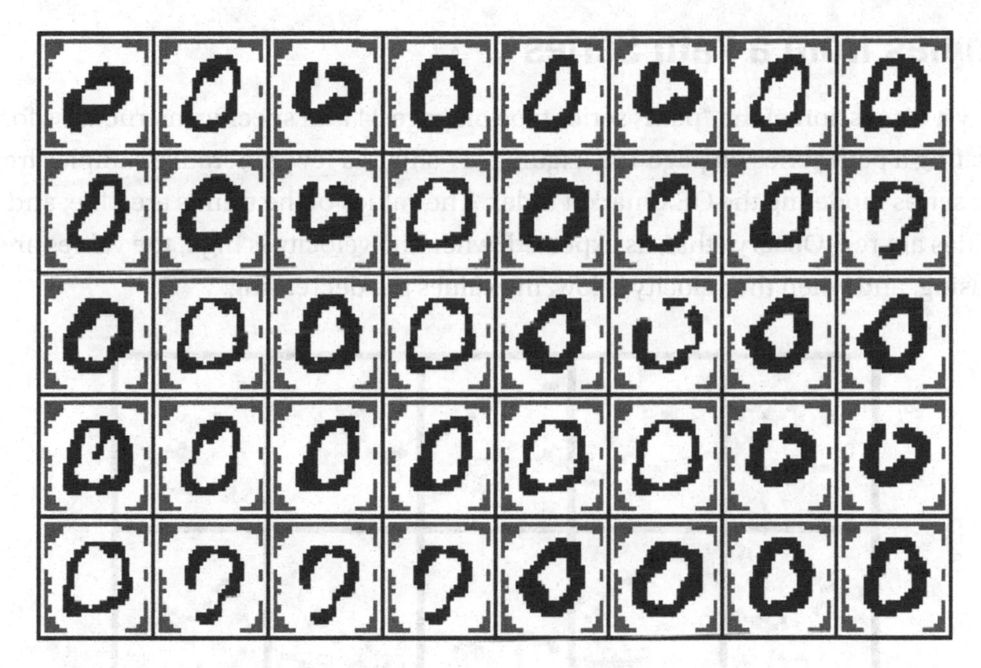

Figure 5-15. *Binary generative samples from MNIST 0 class*

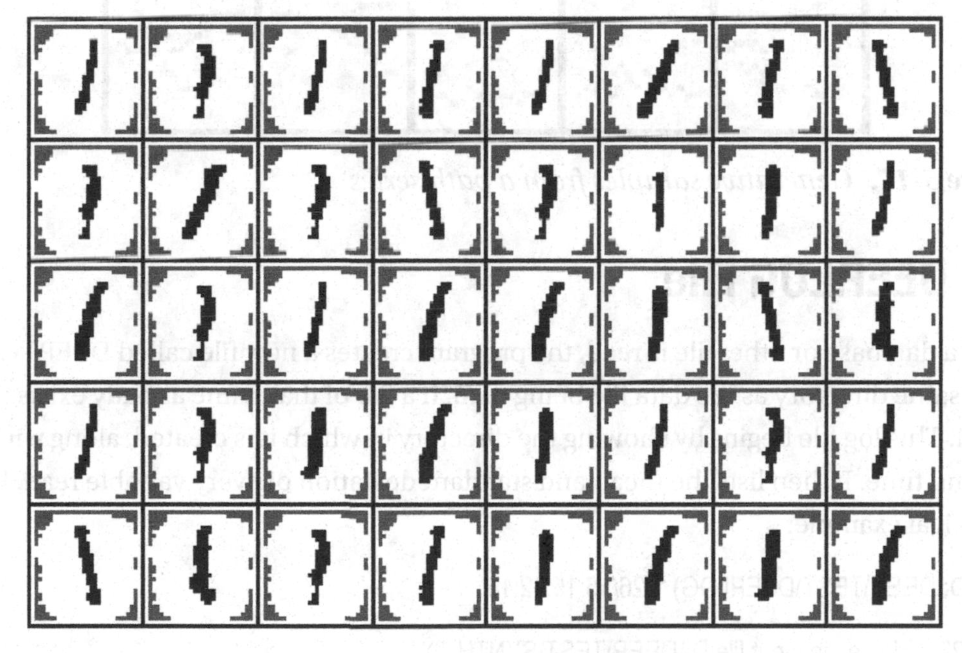

Figure 5-16. *Binary generative samples from MNIST 1 class*

Samples from a Path Series

We saw a discussion of the "path" series type on page 34 and specific instructions for generating a path series on page 199. Figure 5-17 shows a few generative samples from a path series modeling the OEX market index. The values of the trends are blue, and the velocities are red. Observe that, as expected, when the velocity is high, the values are increasing, and when the velocity is low, the values are decreasing.

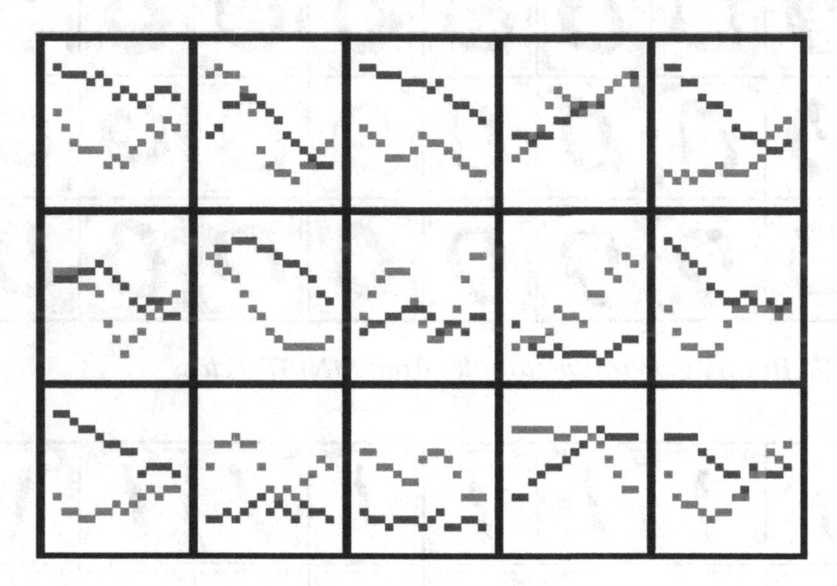

Figure 5-17. *Generative samples from a path series*

The DEEP.LOG File

When a database or other file is read, the program creates a new file called DEEP.LOG in the same directory as the data file being read. If a file of that name already exists, it is erased. This log file begins by showing the directory in which it is created, along with the date and time. It then lists the mean and standard deviation of every variable read. Here is a typical example:

Deep (D:\DEEP\TEST\DEEP.LOG) 1/26/15 15:42:16

Found 23 variables in input file D:\DEEP\TEST\SYNTH.TXT

6304 cases read

Means and standard deviations...

Variable	Mean	StdDev
RAND0	0.00711	0.57541
RAND1	0.01422	0.58043
RAND2	0.01027	0.57694
RAND3	-0.00765	0.58143
RAND4	0.00713	0.57911
RAND5	-0.01166	0.57263
RAND6	-0.00648	0.57742
RAND7	-0.01424	0.58015
RAND8	0.00659	0.57533
RAND9	-0.00366	0.57733

It then shows the architecture of the model, including the unsupervised and supervised sections:

Beginning training a model with the following architecture:

There are 1 unsupervised layers, not including input
Hidden layer has 5 neurons

There are 1 supervised layers, including output

Since there is at least one RBM layer, the training parameters for this layer are listed, as shown here:

Restricted Boltzmann Machine training parameters...
Initial random iterations for starting weights = 50
Number of batches = 24
Markov chain length start = 1
Markov chain length end = 4
Markov chain length rate = 0.0050
Learning rate = 0.05000
Starting momentum = 0.10000

```
  Ending momentum = 0.90000
  Weight penalty = 0.00010
  Sparsity penalty = 0.00100
  Sparsity target = 0.10000
  Increment convergence criterion = 0.00001
  Max epochs with no improvement = 500
  Max epochs = 10000
  Visible layer using mean field, not stochastic
  Inputs will be rescaled to cover a range of 0-1
  Unsupervised section weights will be fine tuned by supervised training
```

The training parameters for the supervised section are also listed:

```
Supervised layer(s) training parameters...
  Initial annealing iterations for starting weights = 100
  Initial random range for starting weights = 1.00000
  Supervised optimization max iterations = 1000
  Supervised optimization convergence tolerance = 0.0000500
  Complete model optimization max iterations = 2000
  Complete model optimization convergence tolerance = 0.0000100
  Weight penalty = 0.00100
```

The results of training the unsupervised layer are printed first:

```
Training unsupervised layer 1
  Initial weight search RMS reconstruction error = 0.27098
  Unsupervised training complete; RMS reconstruction error = 0.31654
```

There is one curious issue in that result. The initial weight search gave a reproduction error of 0.27098, but after real training was done, we see that the reproduction error has increased to 0.31654. How did this happen?

Actually, this is unusual, happening only when the input variables have little or no patterns that the RBM can learn. In this example, the inputs are all random numbers, so there are obviously no patterns. We must remember that the reconstruction error is measured slightly differently during weight initialization and training. In Volume I we see that the initial search reconstruction error is computed in a deterministic manner using mean field approximation in both directions. But during learning we use random sampling of the hidden neuron activations for the reconstruction error. This tends to increase the error somewhat. If the RBM is able to learn real patterns, the difference because of randomization during reconstruction error computation is swamped out by the model's ability to reconstruct authentic patterns. But if there are no patterns to reconstruct, we just get the effect of randomization.

After greedy training of the unsupervised section is complete, the supervised section that follows the unsupervised section is trained. Fine-tuning was selected, so the last step is to tweak the entire model, unsupervised plus supervised sections. Here we see that fine-tuning produces a huge improvement in the criterion, which is negative log likelihood in this example because a classification model was forced.

```
Optimization of supervised section is complete with negative log likelihood = 0.12270
Fine tuning of the entire model is complete with negative log likelihood = 0.02327
```

The targets are listed, and it is noted that the inputs are rescaled 0–1, so the weights that will be printed soon refer to these rescaled values.

```
Trained weights for this model, predicting the following target(s)...
  RAND1
  RAND2
  RAND3

Each raw input has been rescaled 0-1 to cover the min/max range.
Thus, all weights refer to the rescaled value, not the raw value.
```

The weights for the single unsupervised layer are now printed. If there were multiple layers, each set of weights would appear. These weights are *after* fine-tuning.

Weights for unsupervised hidden layer 1

	1	2	3	4	5
Q mean	0.4522	0.4796	0.4556	0.4138	0.4717
skewness	0.1310	0.0700	0.1271	0.2440	0.0618
RAND1	-7.0347	-4.5028	0.9392	-2.5469	1.4879
RAND2	4.7047	-1.7225	-0.5104	7.0462	2.0824
RAND3	2.8726	6.0903	1.6480	-4.7952	-2.6467
RAND4	-0.0131	0.1551	-1.6304	-0.2535	0.3858
RAND5	-0.0032	-0.3523	-0.0453	0.1271	-0.8947
RAND6	-0.0619	-0.1453	-1.8291	-0.1889	-0.2881
BIAS	0.7231	-0.5790	0.8983	-0.6242	0.4237

The model was specified to have five hidden neurons, so we have five columns, one for each. At most ten columns are printed. After each unsupervised layer is trained, the hidden neuron weights are sorted so that the hidden neuron having maximum sum of absolute values becomes the first hidden neuron, and so forth. This way, if we examine the weights to obtain hints about interpretation of features detected, we can focus our efforts on the early columns. However, if fine-tuning is done, as is the case in this example, this sorting can be subverted. This is not a practical problem because fine-tuning almost always largely or entirely destroys the interpretability of weight patterns that were discovered by the RBM.

The *Q mean* row is the mean activation of each hidden neuron, and the *skewness* row is the statistical skewness of the activations. In general, a positive skewness means that the neuron is usually off, and vice versa. These two values are computed *before* fine-tuning; they refer to the actions of the trained RBM before its weights are adjusted by supervised fine-tuning.

We then see the weights that connect the (last and only) unsupervised layer to the (first and only) supervised layer. Also, the final value of the optimization criterion, which we saw earlier, is repeated.

```
Weights for final (output) layer

Target 1 of 3: RAND1

    -9.158017 Unsupervised output 1
    -6.844571 Unsupervised output 2
     0.781757 Unsupervised output 3
    -2.436789 Unsupervised output 4
     2.660202 Unsupervised output 5
     6.449515 CONSTANT

Target 2 of 3: RAND2

     5.160418 Unsupervised output 1
    -1.469535 Unsupervised output 2
    -0.629605 Unsupervised output 3
     9.063721 Unsupervised output 4
     2.708100 Unsupervised output 5
    -8.018184 CONSTANT
```

```
Target 3 of 3: RAND3

    3.467198 Unsupervised output 1
    8.798016 Unsupervised output 2
    1.170767 Unsupervised output 3
   -8.699801 Unsupervised output 4
   -3.433103 Unsupervised output 5
   -2.159271 CONSTANT

Negative log likelihood = 0.02327
```

Last of all the confusion matrix is shown. Usually, when one is training a classifier, the target vector for each case has 1.0 in the position corresponding to the correct class, and 0.0 in all other positions. But this is just a common convention and is not required in DEEP. Instead, whichever target has the maximum value is defined to be the correct class. So, when a model having continuous targets is forced to be a classifier, as is the situation in this example, results are reasonable. In particular, we would expect good classification in this example since all three targets are also present as inputs! Indeed, we see this to be the case.

```
Confusion matrix... Row is true class, column is predicted class
 In each set of three rows for a true class, the first row is the count,
 the second row is the percent for that row (true class)
 and the third row is the percent of the entire dataset.

          1      2      3

  1    2128      3      8
      99.49   0.14   0.37
      33.76   0.05   0.13

  2       9   2088     15
       0.43  98.86   0.71
       0.14  33.12   0.24

  3       8      7   2038
       0.39   0.34  99.27
       0.13   0.11  32.33

Total misclassification = 0.7931 percent
```

In DEEP 2.0, additional confusion matrices will be printed, showing the effect of using thresholds to classify only cases for which the model has varying degrees of certainty. A table of the thresholds for keeping assorted fractions of the training set is also printed.

Predictive Performance Measures

The example output just shown is for a classification model. We now discuss the performance measures that may be seen with a predictive model. The first and most basic statistics are the mean squared error, root-mean-squared error, and R-squared. This output will resemble the following:

```
Mean squared error and R-squared of target(s)...
  MSE of Lead_1 = 0.00007 RMS = 0.00860 RSQ = 0.00396
```

If there are multiple targets, these quantities will be printed separately for each, followed by a pooled value (all targets). It's important to note that these values are computed very differently in DEEP 2.*x* versus DEEP 1.*x*. In version 1, all targets were scaled according to their standard deviations in the training set so that MSE referred to standardized quantities, and these same scale factors were used for any test set(s). There are many advantages to doing it this way, but it led to massive confusion among users, especially when values of R-squared greater than one appeared! For this reason, version 2 now employs the "traditional" approach of avoiding any scaling of the targets, and R-squared for a test set is now based on the variance in that test set, irrespective of the variance in the training set. Of course, negative R-squared values are still possible when a model is anti-predictive (it behaves worse than just guessing the mean). This is fundamental to this test statistic.

The next test statistics are based on a performance figure commonly used in evaluating market trading systems. However, it is broadly applicable (and indeed very useful) in any application in which the goal is not so much to predict an exact value of the target as to predict whether the target will be positive or negative. Moreover, this application must not require that a decision be made; the user has the choice of examining a prediction and then choosing whether to act on this prediction. There is little or no penalty for deliberately ignoring a prediction.

For example, suppose our model predicts the price change of a market in the upcoming day. Based on the prediction we may or may not submit an order to trade. Perhaps the prediction is that the market will rise 1 percent and we take a long position (we buy). Now suppose that we are off by 4 percent in the prediction. Maybe the market actually rises by 5 percent. Or maybe it falls by 3 percent. These two possibilities are equal errors, yet the first error is not a problem at all (!) while the second error is a disaster. So, our "performance penalty" must be similarly asymmetric. Moreover, if the prediction is that the market will rise by 0.0001 percent, we would probably pass on a trade, while a predictied rise of 8 percent will definitely get our attention. Our performance criterion must take this into account.

Suppose we define a threshold for predictions. Given this threshold, we define a *win* as the true value of a target being positive when the prediction equals or exceed the threshold, or the true value being negative when the prediction is less than the threshold. Similarly, we experience a *loss* when the true value is negative when the prediction equals or exceeds the threshold or when the true value is positive while the prediction is less than the threshold.

The condition of the prediction equaling or exceeding the threshold is called *long*, and the condition of the prediction being less than the threshold is called *short*, again in deference to related issues in market trading; an automated trading system would take a long position when the prediction is large, and a short position when the prediction is small (very negative).

A table similar to the one shown on the next page is printed. It is preceded by the *long ratio*, the sum of all positive targets divided by the (absolute) sum of all negative targets. The *net* is the sum of all targets, which of course equals the sum of all positive targets minus the (absolute) sum of all negative targets. The corresponding *short* values are the reciprocal and negative, respectively, of the long values. These serve as baselines for comparison.

```
Total Win vs. Total Loss above and below various fractions
  (For all tested cases, long ratio = 1.1218 (net=3195.592) and s hort ratio = 0.8914
(net=-3195.592)
```

Threshold	Frac Gtr/Eq	Ratio	Net	Frac Less	Ratio	Net
-0.901	0.990	1.1261	3265.0136	0.010	1.2484	69.4217
-0.481	0.950	1.1376	3387.0308	0.050	1.1331	191.4389
-0.279	0.900	1.1489	3455.2076	0.100	1.0937	259.6157
-0.042	0.800	1.1687	3495.6538	0.200	1.0574	300.0619
0.120	0.700	1.1829	3371.6175	0.300	1.0231	176.0256
0.253	0.600	1.1971	3160.0880	0.400	0.9965	-35.5039
0.374	0.500	1.2478	3340.0525	0.500	1.0114	144.4606
0.500	0.400	1.2727	3008.0249	0.600	0.9878	-187.5670
0.642	0.300	1.3077	2608.6799	0.700	0.9680	-586.9120
0.819	0.200	1.3017	1828.5688	0.800	0.9366	-1367.0231
1.072	0.100	1.3382	1106.7696	0.900	0.9166	-2088.8223
1.307	0.050	1.4954	831.8562	0.950	0.9122	-2363.7357
1.789	0.010	1.7472	273.0242	0.990	0.8985	-2922.5677

The rows of this table are computed so as to at least approximately cover preset fractiles of the data distribution, although this may be subverted if there are numerous tied predictions. The leftmost column shows the numerical values of the thresholds corresponding to the preset fractiles. The second column shows the fraction of cases whose prediction equals or exceeds the corresponding threshold. The fifth column is the complement, the fraction of cases whose prediction is less than the threshold.

The third and fourth columns are relevant to the situation of the prediction equaling or exceeding the threshold. The *Ratio* is the sum of wins (positive targets) divided by the (absolute) sum of losses (negative targets). The *Net* is the sum of all targets for theses cases whose predictions equal or exceed the threshold.

Columns six and seven are for cases whose predictions are less than the threshold. For this ratio, wins (the numerator) are negative targets and losses are positive targets.

We see in the line above the table that the long ratio (sum of all positive targets divided by absolute sum of all negative targets) is 1.1218. So, it's no surprise that the ratio in the top row of the third column, 1.1261, is very close to this value because this subset consists of 99 percent of all cases. As we drop to lower rows, corresponding to increasing thresholds, we see the win/loss ratio steadily increasing, until when we reach the point of examining only the highest 1 percent of predictions, the ratio peaks at 1.7472. The net decreases because we have fewer and fewer cases going into the sum.

The reverse happens in the ratio column for cases below the threshold. In the bottom row (99 percent of cases) we have a ratio of 0.8985, close to the entire-set ratio of 0.8914. But by the time we get to the top line, with just 1 percent of the cases having a prediction this small, the win/loss ratio is up to 1.2484.

Not shown here is the fact that this model happens to have a *negative* R- squared! This phenomenon is common in ultra-high-noise situations. Predictive power may be totally nonexistent throughout the majority of the data distribution, but be respectable in the extreme tails. For this reason, charts like the one shown here can be invaluable.

The chart just shown is always computed based on the distribution in the dataset being tested. This generally reveals the most information. However, for out-of-sample testing, this method does have the disadvantage that in real-time applications, such as financial market trading, these thresholds cannot be known in advance. The chart produced by pooling all OOS cases relies on thresholds that would not be known until the entire OOS set has been processed, an obvious impossibility in real time. This does *not* necessarily introduce any bias, optimistic or pessimistic. However, it is an annoyance.

To handle this complaint, the *Test* and *Cross validate* functions print one small additional set of statistics. The three tail thresholds, 0.01, 0.05, and 0.10, both long and short, are preserved for the training set. These same thresholds are then used to compute the win/loss ratios and net sums in any subsequent test set. A typical set of statistics is shown on the next page. For this example, I deliberately used the training data as a test set so that you can compare the results here with those in the prior table. (Please do so.) When different datasets are used, the usual situation, the percent of cases above/below the threshold generally varies from the presets. Here, because the same datasets are used for training and testing, they are virtually identical.

```
Out-of-sample performance at tails, based on training-set thresholds

Target variable Lead_1
0.01    Long n=83 (1.01 Pct) Ratio=1.747 Net=273.024
        Short n=81 (0.99 Pct) Ratio=1.248 Net=69.422

0.05    Long n=411 (5.01 Pct) Ratio=1.495 Net=831.856
        Short n=409 (4.99 Pct) Ratio=1.133 Net=191.439

0.10    Long n=821 (10.01 Pct) Ratio=1.338 Net=1106.770
        Short n=819 (9.99 Pct) Ratio=1.094 Net=259.616
```

Index

A

B

C

T. Masters, *Deep Belief Nets in C++ and CUDA C: Volume 2*, https://doi.org/10.1007/978-1-4842-3646-8

D

E

F

G, H, I, J, K

L

M

N

O, P, Q

R

S

T, U, V, W, X, Y, Z